THIRD EDITION

PRECALCULUS

GUIDED NOTEBOOK

Christopher Schroeder

Project Manager:
Emily Christian

Creative Services Manager:
Trudy Gove

Cover Design:
Lizbeth Mendoza

HAWKES LEARNING

A division of Quant Systems, Inc.

546 Long Point Road
Mount Pleasant, SC 29464

Printed in the United States of America

10 9 8 7 6 5 4 3 2 1

ISBN: 978-1-64277-343-9

Contents

A Message to the Student

Have you ever been working on your homework and know you had some notes somewhere for a problem you were stuck on, but couldn't remember where you put them? Have you ever tried to enter an answer on a homework problem by guessing, but didn't know for sure why you got it wrong (or right)? Have you ever been doing your math homework and wanted to smash your computer with a baseball bat? If you answered "yes" to any of these questions, then this notebook is for you.

In my experiences teaching algebra and trigonometry, I have seen too many students struggle with the material because of a lack of organization and motivation, not because of a lack of ability. This notebook was written with those students in mind, but I believe it can help anyone better understand the concepts of Precalculus.

This notebook is designed to be used in conjunction with the Hawkes Learning Systems' Precalculus courseware. Each section in this book will walk you through three main components: 1) **Learn**; 2) **Watch and Work**; and 3) **Practice**. In the **Learn** section you will be guided through the definitions and examples which are important to understanding the concepts. In **Watch and Work**, you will be given a chance to work some problems similar to videos found on the Hawkes website. Finally, in **Practice**, you will get to work through additional examples that are similar to ones you will find in the *Practice* section of the courseware. At this point, you will have put yourself in a great position to try and *Certify* in that section, with several examples written out and easily accessible in your notebook.

But this Guided Notebook will only help if you are willing to put in the work. So take the time to fill out the sections. Put in the effort to write out the solutions. But whatever you do, don't be afraid to *try*. Don't be afraid of failure! Mathematics can be difficult at times, but I know you have what it takes to succeed.

Chris Schroeder
Morehead State University

Important Information

Name: ______________________________

Class Time: ______________________________

Instructor: ______________________________

Office Hours: ______________________________

Instructor's Phone: ______________________________

Instructor's email: ______________________________

Exam Dates: ______________________________

Final Exam Date: ______________________________

Hawkes Website: learn.hawkeslearning.com

Videos Website: tv.hawkeslearning.com

Notes:

Schedule

Month:						
S	M	T	W	T	F	S
Notes:						

Month:						
S	M	T	W	T	F	S
Notes:						

Month:						
S	**M**	**T**	**W**	**T**	**F**	**S**
Notes:						

Month:						
S	**M**	**T**	**W**	**T**	**F**	**S**
Notes:						

Month:						
S	M	T	W	T	F	S
Notes:						

Month:						
S	M	T	W	T	F	S
Notes:						

Habits for Success

Developing consistent study habits will help you succeed in this course. Writing down what these habits will be prior to implementing them greatly increases the chances that you will utilize these habits effectively. With that in mind, take some time to write out what habits you will develop and use to help you succeed in this course.

My Habits for Success in Precalculus

1. I will begin working on each lesson at least ____________ days before the Certification for that lesson is due. Specifically, when working through a lesson, I will use the following habits:

2. I will begin studying for exams at least ____________ days before taking the exam. Specifically, when studying for an exam, I will use the following habits:

3. I will do my best to attend every class. However, if I have to miss a class, to catch up on the material missed I will use the following habits:

4. If I am stuck on a lesson, I will use the following habits to get help and figure out the material:

Chapter 1

Algebraic Expressions, Equations, and Inequalities

Before we can embark on a journey into algebra, we need to be sure that we know the players and the rules of the game. The players are numbers and expressions, and there are different types of each. The rules tell us how to work with these numbers and expressions. You may already know many of these rules, but it's worth going over them again to make sure you're ready for algebra.

1.1 Real Numbers and Algebraic Expressions

Getting Started

In this section we'll dive into the different types of real numbers, look at the concepts of order and distance, and investigate the terminology and the rules involved in working with algebraic expressions. As you will see, the rules that apply to how we combine and work with numbers also apply to these expressions. These are the first steps in learning the language of algebra and will put us on the path towards fluency in that language. Let's get real.

Topics

1. Common subsets of real numbers
2. The real number line
3. Order on the real number line
4. Set-builder notation and interval notation
5. Basic set operations and Venn diagrams
6. Absolute value and distance on the real number line
7. Components and terminology of algebraic expressions
8. The field properties and their use in algebra

Learn

Build the foundation you need for this section by completing the following.

1. Give an example of:

   (a) a number which is an integer but *not* a natural number. ____________

   (b) a number which is a rational number but *not* an integer. ____________

   (c) two irrational numbers. ____________ ____________

2. The real number corresponding to a given point on the real number line is called the ______________ of that point.

3. List three numbers which are in the set $\{x | x \text{ is an integer and } -1 \le x < 3\}$.

 ____________ ____________ ____________

4. Simplify each of the following set expressions, if possible.

   (a) $(-3, 2] \cup [1, 6] =$ ______________

   (b) $(-3, 2] \cap [1, 6] =$ ______________

5. Given two real numbers a and b, the ______________ between them is defined to be $|a - b|$.

6. **Algebraic expressions** are made up of *constants* and ______________, combined by the operations of addition, subtraction, multiplication, division, exponentiation, and the taking of ____________.

7. Write the terms separately for the algebraic expression $3xy^2 - 7x^2z^3 + 4(xy - z)$.

 ______________ ______________ ______________

8. Write the additive and multiplicative versions of the commutative property.

   Additive:______________ Multiplicative: ______________

Watch and Work

Watch the videos in this section, and then complete the following exercises.

1. Write the set as an interval using interval notation: $\{x | -5 \le x < 3\}$.

2. Evaluate the expression for the given values of the variables:

$$|x + 9y| - (5z + 6) \text{ for } x = -2, y = -1, \text{ and } z = 5$$

Practice

Solve the following problems in the space provided.

1. Evaluate the expression.

$$-|2 - 5|$$

2. Evaluate the expression for the given values of the variables:

$$3\sqrt{x-3} + 7y^3 \text{ for } x = 28 \text{ and } y = -3$$

3. Identify the property that justifies the following statement:

$$-4(6y + 1) = -24y - 4$$

1.2 Properties of Exponents and Radicals

Getting Started

Being able to evaluate expressions with exponents will help you be successful in algebra. We'll look at the properties we use to work with exponents, and then apply those properties to scientific notation and geometric problems. We'll also investigate radicals and their properties. We'll review radical notation and look at some of the methods used to simplify radical expressions. Then we'll study how to combine radical expressions and what to do when we have fractions in an exponent.

Topics

1. Natural number and integer exponents
2. Properties of exponents
3. Scientific notation
4. Working with geometric formulas
5. Radical notation
6. Simplifying and combining radical expressions
7. Rational number exponents

Learn

Build the foundation you need for this section by completing the following.

1. In the expression a^n, a is called the ____________, and n is the ______________.
2. For any real number $a \neq 0$ and for any natural number n, $a^{-n} =$ ____________.
3. Complete the following list of properties of exponents.

(a) $a^n \cdot a^m =$

(b) $\dfrac{a^n}{a^m} =$

(c) $a^{-n} =$

(d) $(a^n)^m =$

(e) $(ab)^n =$

(f) $\left(\dfrac{a}{b}\right)^n =$

4. A number is in scientific notation if it is written in the form ____________ where $1 \leq |a| < 10$ and n is an integer.

5. $\sqrt[n]{a} = b \iff a =$ ____

6. A radical expression is in **simplified form** when:

   (a) The ______________ contains no factor with an exponent greater than or equal to the index of the radical.

   (b) The radicand contains no ______________.

   (c) The denominator, if there is one, contains no ______________.

   (d) The greatest common factor of the ____________ and any exponents occurring in the radicand is 1.

7. If n is a natural number and $\sqrt[n]{a}$ is a real number, then $a^{\frac{1}{n}} =$ _________.

Watch and Work

Watch the videos in this section, and then complete the following exercises.

---

1. Simplify the expression, writing your answer with only positive exponents.

$$\frac{x^4 \cdot x^7}{x^3}$$

2. Simplify the following radical by rationalizing the denominator.

$$\frac{\sqrt{z}+\sqrt{x}}{\sqrt{z}-\sqrt{x}}$$

Practice

Solve the following problems in the space provided.

1. Simplify the expression, writing your answer with only positive exponents.

$$\left(\frac{2a}{b^{-3}}\right)^2$$

2. Evaluate the following expression, using the properties of exponents. Write your answer in standard notation.

$$\frac{(15.3 \times 10^{15})(3 \times 10^{-11})}{(5.1 \times 10^{9})}$$

3. Combine the radical expressions, if possible.

$$\sqrt[3]{-8x^5} + 3x\sqrt[3]{x^2}$$

1.3 Polynomials and Factoring

Getting Started

Polynomials are an important type of algebraic expression and will appear often in this course. With that in mind, let's make sure we are familiar with the terminology of these objects, how to perform operations with them, and how to factor them.

Topics

1. The terminology of polynomial expressions
2. Basic operations with polynomials
3. GCF
4. Factoring by grouping
5. Factoring special binomials
6. Factoring trinomials
7. Factoring expressions containing noninteger rational exponents

Learn

Build the foundation you need for this section by completing the following.

1. Polynomials consisting of a single term are called ____________, those consisting of two terms are ____________, and those consisting of three terms are ____________.
2. In polynomials, **like terms** are terms that have the same ____________ raised to the same ____________.
3. What is the greatest common factor in the polynomial? ____________

$$10a^2b - 6ab + 4a^2b^2$$

4. Factoring by ____________ is the process of grouping the terms of the polynomial in a suitable way to find a common factor among each group.

5. Complete the special binomial factoring formulas.

   (a) $A^2 - B^2 = (\qquad\qquad)(A + B)$

   (b) $A^3 - B^3 = (A - B)(\qquad\qquad\qquad)$

   (c) $A^3 + B^3 = (\qquad\qquad)(A^2 - AB + B^2)$

6. Complete the following steps which are used to factor a trinomial $ax^2 + bx + c$ where $a \neq 1$.

   (a) Multiply a and _____.

   (b) Factor ac into two integers whose sum is _____. If no such factors exist, the trinomial is irreducible over the integers.

   (c) Rewrite b in the trinomial with the sum found in step 2, and _______________. The resulting polynomial of four terms may now be factored by grouping.

7. To factor the expression $x^{\frac{2}{3}} + 3x^{\frac{1}{3}} - 4x^{-\frac{1}{3}}$, what power of x should be factored out first?

Watch and Work

Watch the videos in this section, and then complete the following exercises.

1. Multiply the polynomials.

$$\left(4x^2 + y\right)\left(x^2 - 5y\right)$$

2. Factor the trinomial.

$$12y^2 - 17y + 6$$

Practice

Solve the following problems in the space provided.

1. Subtract the polynomials.

$$\left(-7y^3 + 10 - 9y^4\right) - \left(2 - 4y^3\right)$$

2. Use special factoring patterns to factor the binomial.

$$8a^3 + 27b^6$$

3. Factor the algebraic expression.

$$2x^{-3} + 5x^{-5}$$

1.4 Rational Expressions

Getting Started

Rational expressions are expressions that are able to think clearly and logically. That's a joke. Actually, these are expressions that are made up of a ratio of polynomials. We will learn how to simplify and combine these expressions.

Topics

1. Simplifying rational expressions
2. Combining rational expressions
3. Simplifying complex rational expressions

Learn

Build the foundation you need for this section by completing the following.

1. A **rational expression** is an expression that can be written as a ____________ of two polynomial $\frac{P}{Q}$.

2. Like numerical fractions, in order to add or subtract two rational expressions, a ______________ ______________ must first be found.

3. A ______________ **rational expression** is a fraction in which the numerator or denominator (or both) contains at least one rational expression.

Watch and Work

Watch the videos in this section, and then complete the following exercises.

1. Simplify the rational expression, indicating which real values of the variable must be excluded.

$$\frac{2x^2+5x-3}{x^2-x-12}$$

2. Simplify the complex rational expression.

$$\frac{\frac{a}{b}+\frac{b}{a}}{a+b}$$

Practice

Solve the following problems in the space provided.

1. Simplify the rational expression, indicating which real values of the variable must be excluded.

$$\frac{4x^2 - 13x + 3}{x - 3}$$

2. Add the rational expressions.

$$\frac{x+2}{x-3} + \frac{x^2-4}{x+1}$$

3. Multiply the rational expressions.

$$\frac{3x^2 - +x - 2}{x - 4} \cdot \frac{x^2 - 3x - 4}{x + 1}$$

1.5 Complex Numbers

Getting Started

In this section, we'll study complex numbers. But don't worry, they're actually fairly simple once you get the hang of them. And just like real numbers, we can add, subtract, multiply, and divide complex numbers. In addition, we will encounter equations in this class which do not have a real number solution, but do have a complex number solution. Sounds intriguing...in a complex sort of way.

Topics

1. The imaginary unit i and its properties
2. Basic operations with complex numbers
3. Roots and complex numbers

Learn

Build the foundation you need for this section by completing the following.

1. The imaginary unit i is defined as $i =$ ______. In other words, i has the property that its square is $-1 : i^2 =$ ______.

2. For any two real numbers a and b, the sum ____________ is a **complex number**.

3. Determine the complex conjugate of the complex number $3 - 2i$.
   Conjugate: ______________

4. What is the product of any complex number $a + bi$ and its conjugate $a - bi$?

$$(a + bi)(a - bi) = \_\_\_\_\_\_\_\_\_\_\_\_\_\_$$

Watch and Work

Watch the videos in this section, and then complete the following exercises.

1. Simplify the expression.

$$\sqrt{-20}\sqrt{-2}$$

2. Simplify the quotient.

$$\frac{3-i}{2+4i}$$

Practice

Solve the following problems in the space provided.

1. Simplify the expression.

$$i(10 - 11i)$$

2. Simplify the expression.

$$i^{-43}$$

3. Simplify the expression.

$$\left(3 + \sqrt{-2}\right)^2$$

1.6 Linear Equations in One Variable

Getting Started

Linear equations are one of the most basic types of equations. You probably have solved these types of equations often without even writing anything down. In this section we'll learn the steps necessary to solve any linear equation you may come across. Then we'll look at how this skill may come in handy in a variety of applications.

Topics

1. Equations and the meaning of solutions
2. Solving linear equations
3. Solving linear absolute value equations
4. Solving linear equations for one variable
5. Applications of linear equations

Learn

Build the foundation you need for this section by completing the following.

1. Give an example of each of the following types of equations.

   Identity:______________ Contradiction:______________

   Conditional:______________

2. A **linear equation in one variable**, say the variable x, is an equation that can be transformed into the form __________________, where a and b are real numbers and $a \neq 0$.

3. The absolute value equation $|ax + b| = c$ means $ax + b = c$ or

   ______________ $= c$.

4. **Solving for a variable** means to transform the equation into an equivalent one in which the specified variable is ____________ on one side of the equation.

5. In the distance formula $d = rt$, what does r stand for?

Watch and Work

Watch the videos in this section, and then complete the following exercises.

---

1. Solve the linear equation.

$$0.5t + 1.4 = 3t$$

2. Solve the absolute value equation.

$$|3x - 7| = 13$$

Practice

Solve the following problems in the space provided.

1. Solve the linear equation.

$$2t + 1 = 3(t + 2) - 5$$

2. Solve the linear equation.

$$\frac{2y + 3}{4} + \frac{3}{2} = \frac{3y - 1}{2}$$

3. Solve the absolute value equation.

$$|x + 5| = |x - 4|$$

1.7 Linear Inequalities in One Variable

Getting Started

Oftentimes, we want to find all of the numbers that make a particular inequality true. Since there will usually be infinitely many solutions to such a problem, it would take quite a while to write them all down. Instead, we'll represent our answers either graphically or using interval notation.

Topics

1. Solving linear inequalities
2. Solving double linear inequalities
3. Solving linear absolute value inequalities
4. Applications of linear inequalities

Learn

Build the foundation you need for this section by completing the following.

1. Fill in the blanks with $<$ or $>$.

   (a) If $A < B$ and $D > 0$, the $A \cdot D$______$B \cdot D$.

   (b) If $A < B$ and $D < 0$, the $A \cdot D$______$B \cdot D$.

2. A **compound inequality** is a statement containing two inequality symbols and can be interpreted as two distinct inequalities joined by the word "__________."

3. To solve the absolute value inequality $|2x + 6| > 3$, what two independent inequalities would you solve?

   __________________ or __________________

4. Fill in the table for the given phrase.

Phrase	Can x be less than y?	Can x equal y?	Can x be greater than y?	Inequality
"y is at least as large as x"				

Watch and Work

Watch the videos in this section, and then complete the following exercises.

1. Solve the absolute value inequality. Describe the solution set using interval notation and by graphing.

$$2|a+4| \leq 16$$

2. Solve the inequality problem. Describe the solution set using interval notation and by graphing.

$$-3(x+2) \leq 12 \text{ or } 15 + x < 17$$

Practice

Solve the following problems in the space provided.

1. Solve the double inequality. Describe the solution set using interval notation and by graphing.

$$\frac{6}{8} < \frac{y+2}{4} < \frac{18}{8}$$

2. Solve the absolute value inequality. Describe the solution set using interval notation and by graphing.

$$-3 + |z - 5| \geq -1$$

3. In a class in which the final course grade depends entirely on the average of four equally weighted 100-point tests, Jason has scored 87, 91, and 83 on the first three. What range of scores on the fourth test will give Jason a B for the semester (an average between 80 and 89, inclusive)? Assume that all test scores have a non-negative integer value.

1.8 Polynomial and Polynomial-Like Equations in One Variable

Getting Started

Quadratic equations are one of the most common types of equations. There are numerous methods that can be used to solve quadratic equations, including the quadratic formula, which can solve any quadratic equation in the world. Talk about raw power. But as you may have guessed, there are numerous equations with a degree of more than 2 that can be solved. Unfortunately, these equations usually also have a higher degree of difficulty. In this section, we will look at some of the methods that can be used to solve different types of these polynomial equations.

Topics

1. Solving quadratic equations by factoring
2. Solving "perfect square" quadratic equations
3. Solving quadratic equations by completing the square
4. The quadratic formula
5. Applications of quadratic equations
6. Solving quadratic-like equations
7. Solving general polynomial equations by factoring
8. Solving polynomial-like equations by factoring

Learn

Build the foundation you need for this section by completing the following.

1. A **quadratic equation in one variable**, say the variable x, is an equation that can be transformed into the form ________________, where a, b, and c are real numbers and $a \neq 0$.

2. If A is an algebraic expression and if c is a constant, the equation $A^2 = c$ means $A = \sqrt{c}$ or

$$A = \_\_\_\_\_.$$

3. If you wanted to solve the equation $x^2 + 6x - 4 = 0$ by **completing the square**, what would the equation look like after the first step?

4. According to the **quadratic formula**, the solutions of the equation $ax^2 + bx + c = 0$ are

$$x = \qquad\qquad\qquad\qquad .$$

5. If a ball is thrown upward with a velocity of 20 ft/s from a height of 42 ft, what values would be substituted for v_0 and h_0 in the formula $h = -\frac{1}{2}gt^2 + v_0t + h_0$?

$$v_0 = \_\_\_\_\_\_\_\_, \quad h_0 = \_\_\_\_\_\_\_\_$$

6. An equation is quadratic-like, or quadratic in form, if it can be written in the form ______________ where $a, b,$ and c are constants, $a \neq 0$, and A is an algebraic expression.

7. In the quadratic-like equation below, determine the value of A so that the equation can be written as $aA^2 + bA + c = 0$.

$$(x^3 + x)^2 - (x^3 + x) - 12 = 0 \qquad\qquad A = \_\_\_\_\_\_\_\_\_\_\_\_$$

8. To solve $8x^3 - 27 = 0$ by factoring, we would write the left side of the equation as a difference of two cubes. What are the two things being cubed?

$$(\qquad)^3 - (\qquad)^3 = 0$$

9. In the equation $x^{5/2} + 2x^{3/2} - 8x^{1/2} = 0$, what power of x can be factored out on the left side of the equation?

Watch and Work

Watch the videos in this section, and then complete the following exercises.

1. Solve the equation $x^2 + 6x + 9 = 64$ by the square root method.

2. Solve the polynomial equation by factoring.

$$2x^3 + x^2 + 2x + 1 = 0$$

Practice

Solve the following problems in the space provided.

1. Solve the quadratic equation using the quadratic formula.

$$2x^2 - 4x - 3 = 0$$

2. How long would it take for a stone dropped from the top of a 196-foot building to hit the ground?

3. Solve the polynomial equation by factoring.

$$8x^3 - 1 = 0$$

1.9 Rational and Radical Equations in One Variable

Getting Started

Equations that involve rational expressions are, not surprisingly, called rational equations. In this section, you'll learn how to use your rational mind to solve these rational equations. Some equations are a little more out there than others. These are called radical equations, and there is a very specific set of steps we can use to solve them, which we'll look at in this section.

Topics

1. Solving rational equations
2. Applications of rational equations
3. Solving radical equations
4. Solving equations with positive rational exponents
5. Solving equations for one variable

Learn

Build the foundation you need for this section by completing the following.

---

1. A **rational equation** is an equation that contains at least one rational expression, while any nonrational expressions are _______________.

2. In the following rational equation, what is the LCD that should be multiplied by both sides of the equation?

$$\frac{x}{x-1} + \frac{2}{x-3} = \frac{2}{x^2 - 4x + 3} \qquad \text{LCD} = \underline{\hspace{4cm}}$$

3. There are two keys to solving a work-rate problem:

   (1) The rate of work is the _______________ of the time needed to complete the task.

   (2) Rates of work are "_______________."

4. Fill in the blanks for the procedure for solving radical equations.

   Step 1: Begin by ____________ the radical expression on one side of the equation. If there is more than one radical expression, choose one to isolate on one side.

   Step 2: Raise both sides of the equation by the power necessary to "__________" the isolated radical. That is, if the radical is an nth root, raise both sides to the ________ power.

   Step 3: If any radical expressions remain, simplify the equation if possible and then repeat steps 1 and 2 until the result is a _______________ equation. When a polynomial equation has been obtained, ____________ the equation using polynomial methods.

   Step 4: ____________ your solutions in the original equation! Any extraneous solutions must be discarded.

5. Isolate the radical in the following equation.

   $\sqrt{x-3}+4=x$ $\qquad\qquad$ $\sqrt{x-3}=$ ____________

6. To what power should we raise both sides of the following equation?

   $\left(2x^2-4x+12\right)^{1/5}=3$ $\qquad$ ______ power

Watch and Work

Watch the videos in this section, and then complete the following exercises.

---

1. Solve the rational equation.

$$\frac{1}{y-2}+\frac{1}{y-1}=\frac{3}{y-1}$$

2. Solve the radical equation.

$$\sqrt{50 + 7s} - s = 8$$

Practice

Solve the following problems in the space provided.

1. Solve the rational equation.

$$\frac{4}{x-2} + \frac{2}{x-1} = 1$$

2. If Angela were to clean her bathroom alone, it would take 2 hours. Her sister Kendra could do the job in 4 hours. How long would it take them to clean the bathroom working together?

3. Solve the radical equation.

$$\sqrt[3]{6y^2 + 27y} = \sqrt[3]{2y^2 + 40}$$

Chapter 2

Equations and Inequalities in Two Variables

In the previous chapter, we studied equations and inequalities that contained only one variable. We were able to list our solutions as a single number or, in the case of inequalities, as a section of the real number line. In this chapter we turn our attention to equations and inequalities of *two* variables. In the case of equations, our solutions will consist of ordered pairs. In the case of inequalities, our solutions will be sections of a two-dimensional coordinate system.

2.1 The Cartesian Coordinate System

Getting Started

You've probably worked with the Cartesian coordinate system before, but in this section we will look at the basic properties of this system and familiarize ourselves with a way to find the distance between two points as well as the midpoint of two points.

Topics

1. The Cartesian coordinate system
2. Graphing equations
3. The distance and midpoint formulas

Learn

Build the foundation you need for this section by completing the following.

1. The **Cartesian coordinate system** consists of two ______________ real number lines intersecting at the ______ point of each line.

2. In which quadrant would each of the following points be plotted?

   (a) $(-3, 2)$ Quadrant ______

   (b) $(3, -2)$ Quadrant ______

   (c) $(3, 2)$ Quadrant ______

   (d) $(-3, -2)$ Quadrant ______

3. The **graph of an equation** is a plot in the Cartesian plane of all of the ______________ __________ that make up the solution set of the equation.

4. The distance between two points, (x_1, y_1) and (x_2, y_2) is given by

$$d = \underline{\hspace{5cm}}.$$

5. The midpoint between two points, (x_1, y_1) and (x_2, y_2) is given by

$$(\underline{\hspace{1.5cm}}, \underline{\hspace{1.5cm}}).$$

Watch and Work

Watch the videos in this section, and then complete the following exercises.

1. Fill in the table so that the ordered pairs are solutions of the equation $y = x^2 - x - 2$.

x	0		2
y		1	

2. Find the distance between $(4, 1)$ and $(-2, -1)$.

Practice

Solve the following problems in the space provided.

1. Consider the following pair of points: $(8, -6)$ and $(6, -3)$.

   (a) Determine the distance between the two points.

   (b) Determine the midpoint of the line segment joining the pair of points.

2. Find the perimeter of the triangle whose vertices are the following specified points in the plane.

$$(-9, 2), (-7, -1) \text{ and } (3, 1)$$

3. Consider the following points: $(9, 4), (-4, 2)$ and $(-4, 4)$. Determine whether or not the given points form a right triangle. If the triangle is **not** a right triangle, determine if it is isosceles or scalene.

2.2 Circles

Getting Started

Another type of two-variable equation is the equation of a circle. In order to graph a circle, we need to know two things: the center of the circle and the radius of the circle. We will look at forms of equations of circles that allow us to quickly identify these two important items–without any circular logic.

Topics

1. The standard form of the equation of a circle
2. Graphing circles

Learn

Build the foundation you need for this section by completing the following.

1. The **standard form** of the equation for a circle of radius r with center (h, k) is ____________________.

2. Determine the center and radius of the circle with equation $(x - 3)^2 + (y + 4)^2 = 16$.

   Center:(______, ______), Radius:______

3. Fill in the blanks to complete the square for the following equation of a circle.

$$\begin{aligned} x^2 + y^2 + 4x - 6y &= 3 \\ (x^2 + 4x + \_\_\_\_) + (y^2 - 6y + \_\_\_\_) &= 3 + \_\_\_\_ + \_\_\_\_ \\ (x + \_\_\_\_)^2 + (y - \_\_\_\_)^2 &= \_\_\_\_ \end{aligned}$$

Watch and Work

Watch the videos in this section, and then complete the following exercises.

1. Find the equation of the circle with diameter endpoints $(3, 5)$ and $(-1, -3)$.

2. Find the center and radius of the circle with equation $x^2 + y^2 - 2x + 10y = -1$.

Practice

Solve the following problems in the space provided.

1. Consider the equation below. Find the center and radius of this circle.

$$\left(x-\frac{3}{4}\right)^2+\left(y+\frac{2}{7}\right)^2=\frac{16}{25}$$

2. Find the standard form of the equation for the circle with radius $\sqrt{5}$ and center $(-3,6)$.

3. Find the standard form of the equation for the circle with center $(5, -2)$ and tangent to the x-axis.

2.3 Linear Equations in Two Variables

Getting Started

The solution set of a linear equation in two variables is a set of ordered pairs whose graph is a line in the Cartesian plane. Our job in this section will be to recognize these types of equations and what the graphs of the solutions will look like. We'll also look at the special cases of horizontal and vertical lines. Let's stay on the straight and narrow.

Topics

1. Recognizing linear equations in two variables
2. Intercepts of the coordinate axes
3. Horizontal and vertical lines

Learn

Build the foundation you need for this section by completing the following.

1. A **linear equation in two variables**, say the variables x and y, is an equation that can be written in the form __________________, where a, b, and c are constants and a and b are not both zero. This form of such an equation is called the **standard form**.

2. To find the x-intercept of a linear equation, we let $y =$ _____ and solve for x. To find the y-intercept of a linear equation, we let $x =$ _____ and solve for y.

3. An equation of the form $ax = c$ represents a _______________ line with an x-intercept of ______.

4. An equation of the form $by = c$ represents a _______________ line with a y-intercept of ______.

Watch and Work

Watch the videos in this section, and then complete the following exercises.

1. Determine if the equation is linear.

$$\frac{6}{x} - \frac{5}{y} = 2xy$$

2. Find the x- and y-intercepts of the equation $3y - 5x = -4(1 + x)$.

Practice

Solve the following problems in the space provided.

1. Determine if the following equation is linear.

$$\frac{12x - y}{5} + 8y = 2$$

2. Determine if the following equation is linear. If the equation is linear, then write it in standard form: $ax + by = c$.

$$(10 + y)^2 - y^2 = 11x + 2$$

3. Determine the x- and y-intercepts of the following linear equation, if possible, and then sketch a graph of the equation.

$$11x + 8 = -11y + 8$$

2.4 Slope and Forms of Linear Equations

Getting Started

Linear equations can appear in a variety of forms. Depending on what you want to do with the equation, some forms are more useful than others. In this section, we'll look at some of the different forms of linear equations and how they can be used to get information about the graph of the corresponding line.

Topics

1. The slope of a line
2. The slope-intercept form of the equation of a line
3. The point-slope form of the equation of a line

Learn

Build the foundation you need for this section by completing the following.

---

1. Given a line L through the points (x_1, y_1) and (x_2, y_2), the **slope** of the line L is the ratio ______________.

2. The only lines with slope equal to 0 are ______________ lines.

3. The only lines for which the slope is undefined are ______________ lines.

4. The **slope-intercept form** of a line is an equation of the form ________________ where ______ is the slope of the line, and the y-intercept of the line is $(0,$ _____$)$.

5. The **point-slope form** of the equation for the line passing through the point (x_1, y_1) with slope m is ________________.

Watch and Work

Watch the videos in this section, and then complete the following exercises.

1. Find an equation in slope-intercept form of the line through $(0, 3)$ with slope -4.

2. Find an equation of the line, in slope-intercept form, that goes through the points $(-2, 3)$ and $(4, -1)$.

Practice

Solve the following problems in the space provided.

1. Determine the slope of the line passing through the specified points.

$$(0, 2) \text{ and } (3, -5)$$

2. Determine the slope of the line defined by the equation.

$$4x - y = 22$$

3. Find the equation, in standard form, of the line passing through the specified points.

$$(-3, 1) \text{ and } (1, 3)$$

2.5 Parallel and Perpendicular Lines

Getting Started

When we encounter a pair of linear equations, one of the things we will be interested in is whether or not these lines intersect. If they don't intersect, the two lines are parallel. If they do intersect, we then may want to determine if they do so at a right angle. If this is the case, we say that the two lines are perpendicular.

Topics

1. Slopes of parallel lines
2. Slopes of perpendicular lines

Learn

Build the foundation you need for this section by completing the following.

---

1. Two nonvertical lines with slopes m_1 and m_2 are parallel if and only if ________________.

2. If m_1 and m_2 represent the slopes of two perpendicular lines, neither of which is vertical, then

$$m_1 = \_\_\_\_\_\_\_\_\_\_\_\_.$$

3. If one of two perpendicular lines is vertical, the other is ________________, and the slopes are, respectively, ________________ and zero.

4. Which of the following lines is perpendicular to the line $y = -5x + 3$? Circle the correct answer.

$$y = \frac{1}{5}x - 2 \qquad y = -5x - 2 \qquad y = -\frac{1}{5}x - 2$$

Watch and Work

Watch the videos in this section, and then complete the following exercises.

1. Are the following lines parallel, perpendicular, or neither?

$$5-(4y+3x)=5(x-y) \quad \text{and} \quad y+4=3+8x$$

2. Find the equation of the line through $(8,5)$, perpendicular to $\dfrac{3x-y}{4}=\dfrac{4x-5}{2}$.

Practice

Solve the following problems in the space provided.

1. Find the equation, in slope-intercept form, for the line that is parallel to the following line and passes through the point $(-4, 12)$.

$$1 - \frac{3y + 6x}{3} = 3$$

2. Find the equation, in slope-intercept form, for the line that is perpendicular to the following line and passes through the point $(3, -3)$.

$$7x + 5y = 4y - 4$$

3. Determine if the two lines are parallel, perpendicular, or neither.

$$\frac{7x - 3y}{4} = x + 2 \quad \text{and} \quad -3y - 2x = x + 1$$

2.6 Linear Inequalities in Two Variables

Getting Started

When we solved linear inequalities in one variable, one of the ways we represented our solution was by shading in a portion of the real-number line. To solve linear inequalities in two variables, we will need to represent our solutions graphically by shading in a portion of the Cartesian plane. In this section, we will look at the methods used to determine what part of the plane to shade in.

Topics

1. Graphing linear inequalities
2. Graphing linear inequalities joined by "and" or "or"
3. Graphing linear absolute value inequalities

Learn

Build the foundation you need for this section by completing the following.

1. A linear inequality in the two variables x and y is an inequality that can be written in the form

   $\underline{ax + by < c,}$ ________________, ________________, or ________________,

   where $a, b,$ and c are constants and a and b are not both 0.

2. In solving linear inequalities in two variables, we make the boundary line ______________ if the inequality symbol is $\leq$ or $\geq$ and ________________ if the symbol is $<$ or $>$.

3. The union of two sets A and B, denoted $A \cup B$, is the set containing all elements that are in both A _____ B, while the intersection of two sets A and B, denoted $A \cap B$, is the set containing all elements that are in both A _____ B.

4. What two inequalities must be solved to solve the absolute value inequality $|x + 3| > 1$?

   __________________ or __________________

Watch and Work

Watch the videos in this section, and then complete the following exercises.

1. Solve the linear inequality by graphing its solution set.

$$3x + 3 < -3y + 1$$

2. Graph the solution set that satisfies the linear absolute value inequality.

$$|4x - 2y| > 8$$

Practice

Solve the following problems in the space provided.

1. Graph the solution set that satisfies the inequalities.

$$2x + 2y < -4 \text{ and } x \geq 5$$

2. Graph the solution set that satisfies the inequalities.

$$3y < -4x + 18 \text{ or } 4y \geq -3x + 18$$

3. Graph the solution set that satisfies the linear absolute value inequality.

$$|4x + 8| \leq 18 \quad \text{or} \quad |4y + 8| \leq 18$$

Chapter 3

Relations, Functions, and Their Graphs

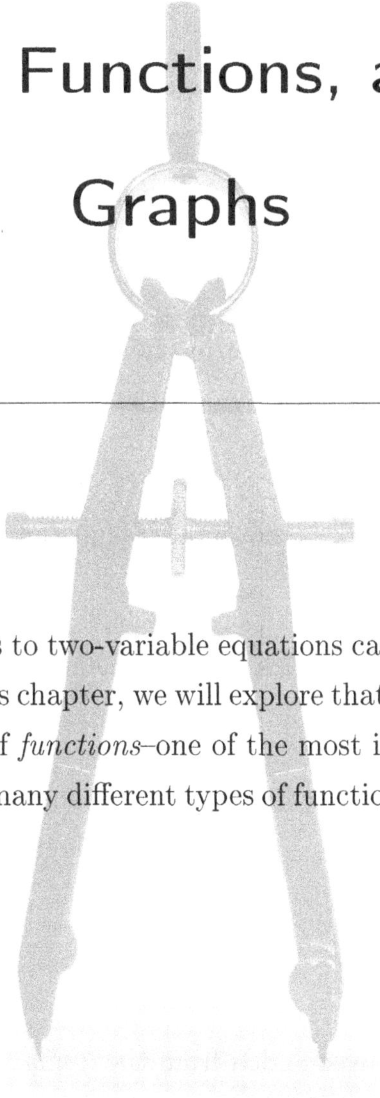

We've seen that the solutions to two-variable equations can be represented as a graph in the Cartesian plane. In this chapter, we will explore that idea further. In particular, we will introduce the topic of *functions*–one of the most important concepts in all of mathematics. We will learn many different types of functions and familiarize ourselves with their respective graphs.

3.1 Relations and Functions

Getting Started

Before we can understand functions, we must first look at *relations.* As we'll see, functions are a type of relation. Recognizing which relations are also functions is one of the main things we'll learn in this section.

Topics

1. Relations, domain, and range
2. Functions and the vertical line test
3. Function notation and function evaluation
4. Implied domain of a function

Learn

Build the foundation you need for this section by completing the following.

1. A **relation** is a set of ______________ pairs. The **domain** of a relation is the set of all the ______________ coordinates, and the **range** of a relation is the set of all second ______________.

2. A **function** is a relation in which every element of the domain is paired with ______________ one element of the range.

3. If a relation can be graphed in the Cartesian plane, the relation is a function if and only if no ______________ line passes through the graph more than once.

4. For the function $y = -3x + 1$, x is called the ______________ variable and y is called the ______________ variable.

5. Which value for x must be excluded from the domain of the function?

$$f(x) = \frac{x^2 - 3x}{x - 2} \qquad x \neq \_\_\_\_\_\_\_\_\_\_$$

Watch and Work

Watch the videos in this section, and then complete the following exercises.

1. Rewrite the relation as a function of x. Then, evaluate the function at $x = -1$.

$$\frac{2y + x}{3} = 4$$

2. Determine the implied domain of the function.

$$f(x) = \sqrt{3 + x} - 2$$

Practice

Solve the following problems in the space provided.

1. Given the relation $3y - 5x = -3$, describe the domain and range of the relation and determine if the relation is a function.

2. Given the function $f(x) = \sqrt{x+9} - 3$, find $f(x-1)$.

3. Given the following function, determine the difference quotient, $\dfrac{f(x+h)-f(x)}{h}$.

$$f(x) = x^2 - 3x + 2$$

3.2 Linear Functions

Getting Started

We've already looked at many of the properties of linear functions in Chapter 3, although we were not considering them to be functions at that time. If we can solve a linear equation in two variables for the variable y, then we can think of our equation as a linear function. These types of functions have many uses in mathematics and beyond, including linear regression.

Topics

1. Linear functions and their graphs
2. Linear regression

Learn

Build the foundation you need for this section by completing the following.

1. A **linear function** f of one variable, say the variable x, is any function that can be written in the form ________________, where m and b are real numbers.

2. The graph of a linear function is a ______________ __________ with slope m and y-intercept $(0, b)$.

3. Linear regression gives us the slope-intercept form for the line whose graph minimizes the ______ or the ____________ of the deviations between the line and the actual data points. For this reason, the specific process we use is known as the ____________-____________ method.

4. The **Pearson correlation coefficient** r always satisfies

$$\_\_\_\_\_ \leq r \leq \_\_\_\_\_.$$

Watch and Work

Watch the videos in this section, and then complete the following exercises.

1. Graph the linear function.

$$r(x) = 2(4 - 3x) + 5x$$

2. Find a formula for the linear function depicted in the graph.

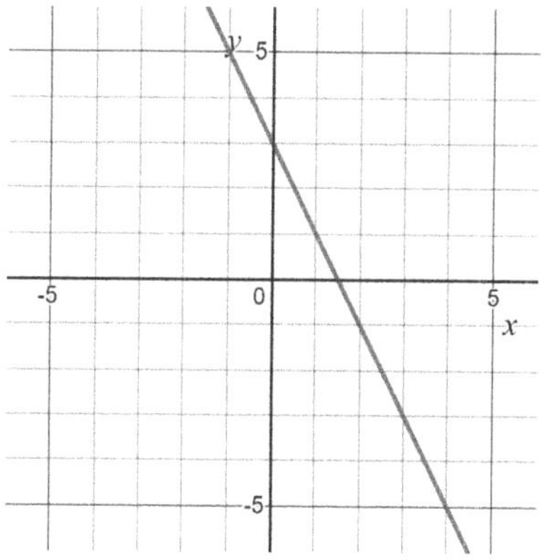

Practice

Solve the following problems in the space provided.

1. Graph the linear function.

$$a(x) = (4x - 18) - (-14 + 6x)$$

2. Find a formula for the linear function depicted in the graph.

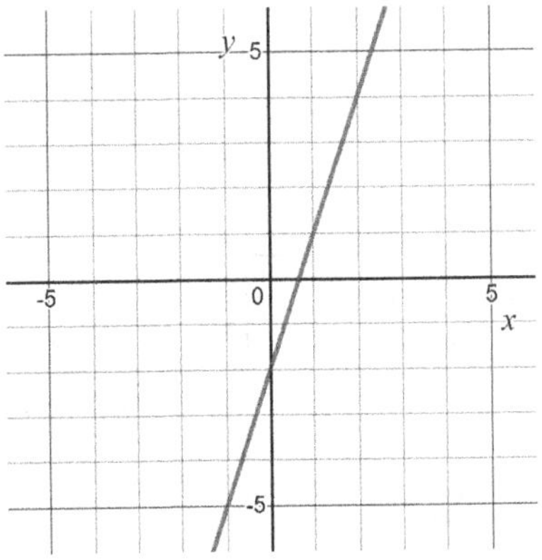

3. For the given points, use linear regression to find the line of best fit, and find the Pearson correlation coefficient, r.

$$(-1, 5), (0, 3), (1, 2), (2, 0), (3, -1)$$

3.3 Quadratic Functions

Getting Started

Quadratic functions are very important and common. They have graphs which are parabolas. One of the skills we'll learn is how to find the vertex of a parabola for a given quadratic function. This will come in handy when using these functions to find the maximum or minimum values for real-world situations.

Topics

1. Quadratic functions and their graphs
2. Quadratic regression
3. Maximization/minimization problems

Learn

Build the foundation you need for this section by completing the following.

1. A **quadratic, or second-degree, function** f in the variable x, is any function that can be written in the form ________________, where a, b, and c are real numbers and $a \neq 0$.

2. The graph of the function $g(x) = a(x-h)^2 + k$, where a, h, and k are real numbers and $a \neq 0$, is a parabola whose vertex is at __________.

3. If we think that a parabola reflects the behavior of points, the least-squares method can be used can be used to find a quadratic function that best fits the points. This process is called ____________ ____________.

4. The vertex will be the ____________ value of a function (if the parabola opens upward) or the ____________ value (if the parabola opens downward).

Watch and Work

Watch the videos in this section, and then complete the following exercises.

1. Graph the quadratic function, accurately locating the vertices and x-intercepts (if any).

$$f(x) = (x + 2)^2 - 4$$

2. Find a formula for the corresponding quadratic function for the parabolic graph.

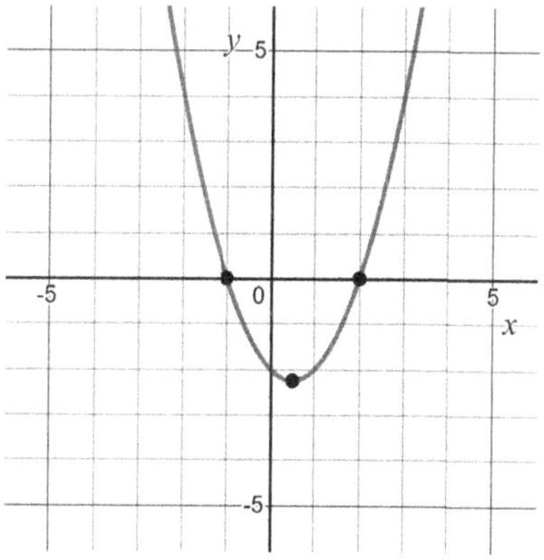

Practice

Solve the following problems in the space provided.

1. Graph the quadratic function, accurately locating the vertices and x-intercepts (if any).

$$f(x) = x^2 - 2x - 8$$

2. Given the points graphed in the figure, use quadratic regression to find and graph the quadratic function of best fit.

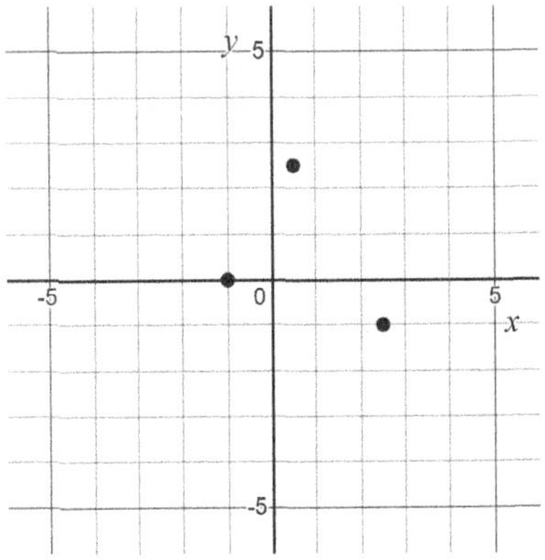

3. Among all rectangles that have a perimeter of 36, find the dimensions of the one whose area is largest.

3.4 Other Common Functions

Getting Started

Of course, there are *many* more types of functions besides linear and quadratic functions. In this section, we'll look at the properties and graphs of some other types of functions which are common in mathematics. We'll also investigate functions whose formula changes depending on which number is put into it. Sounds like uncommon fun.

Topics

1. Power functions of the form ax^n
2. Power functions of the form ax^{-n}
3. Power functions of the form $ax^{\frac{1}{n}}$
4. The absolute value function
5. The greatest integer function
6. Piecewise-defined functions

Learn

Build the foundation you need for this section by completing the following.

1. For functions of the form $f(x) = ax^n$, what is the domain and range when

   (a) n is even? Domain:____________ Range:____________

   (b) n is odd? Domain:____________ Range:____________

2. For functions of the form $f(x) = \frac{a}{x^n}$, what is the domain and range when

   (a) n is even? Domain:____________ Range:____________

   (b) n is odd? Domain:____________ Range:____________

3. For functions of the form $f(x) = ax^{1/n}$, what is the domain and range when

   (a) n is even? Domain:________________ Range:________________

   (b) n is odd? Domain:________________ Range:________________

4. The absolute value function, $f(x) = |x|$, can be thought of as a piecewise-function using the definition of absolute value:

$$|x| = \begin{cases} \underline{\qquad\qquad} & \text{if } x \geq 0 \\ \underline{\qquad\qquad} & \text{if } x < 0 \end{cases}$$

5. The greatest integer of x, $[[x]]$, is the largest integer ______________ ______________ or equal to x.

6. A **piecewise-defined function** is a function defined in terms of two or more formulas, each valid for its own unique portion of the ________ ______________ ________.

Watch and Work

Watch the videos in this section, and then complete the following exercises.

1. Sketch a graph of the function. Find and plot at least two ordered pairs.

$$t(x) = 6|x|$$

2. From the list of functions, write the one that corresponds with each of the following graphs.

$$f(x) = x^4, \quad f(x) = \frac{1}{x^4}, \quad f(x) = \sqrt[4]{x}, \quad f(x) = 4|x|$$

(a)

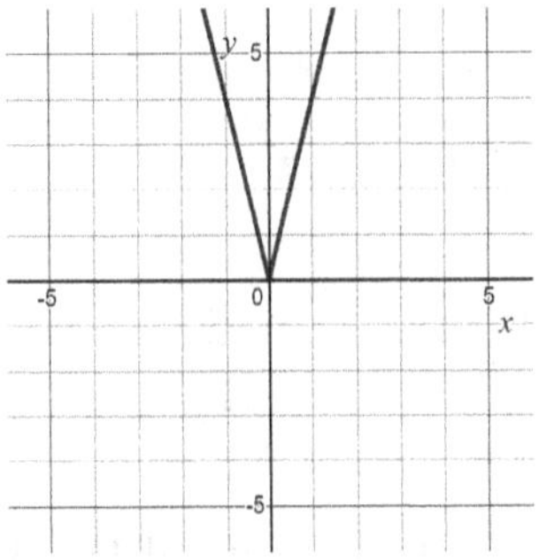

____

(b)

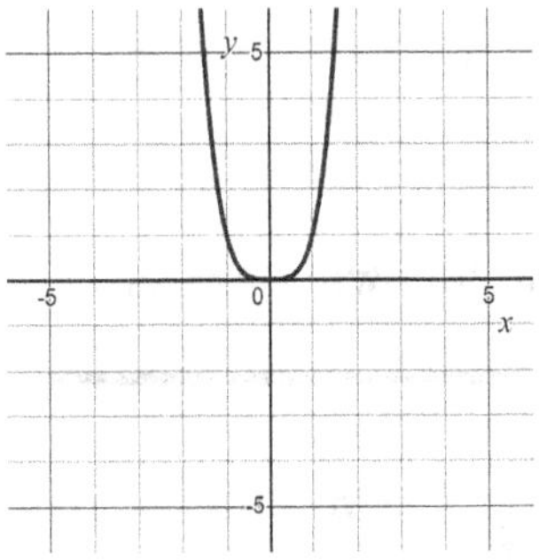

____

(c)

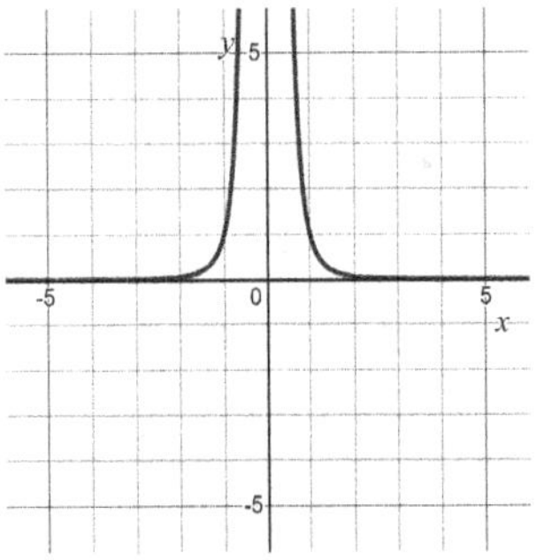

____

(d)

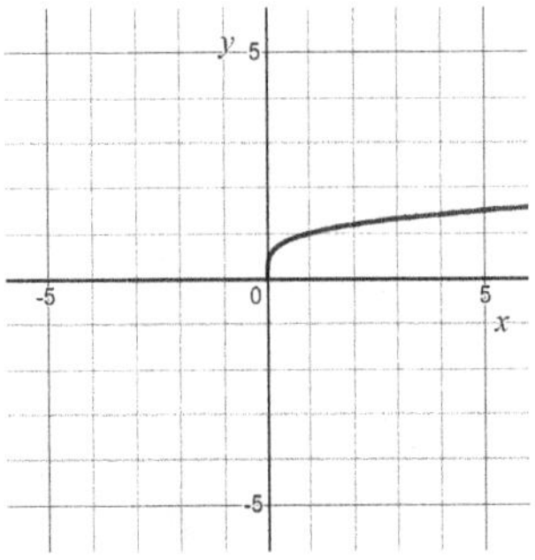

____

Practice

Solve the following problems in the space provided.

1. Sketch a graph of the function. Find and plot at least two ordered pairs.

$$f(x) = \frac{9}{8x^3}$$

2. Sketch a graph of the function. Find and plot at least two ordered pairs.

$$g(x) = -2[[x]]$$

3. Consider the piecewise-defined function $\begin{cases} 3x+1 & \text{if } x \leq -4 \\ \frac{2}{3}x - 7 & \text{if } x > -4 \end{cases}$.

   (a) Evaluate the function at $x = -4$.

   (b) Evaluate the function at $x = -8$.

   (c) Evaluate the function at $x = 0$.

3.5 Variation and Multivariable Functions

Getting Started

There are a lot of instances of a quantity changing as the result of another quantity changing. This is called **variation**. If a quantity increasing causes another quantity to increase, we call it **direct variation**. If a quantity increasing causes another quantity to decrease, we call it **inverse variation**. In other words, there are variations of variation. Combining those we will also look at functions with more than one input variable using variation.

Topics

1. Direct variation
2. Inverse variation
3. Joint variation
4. Multivariable functions

Learn

Build the foundation you need for this section by completing the following.

1. We say that y **varies directly as** the nth power of x if there is a nonzero constant k such that $y =$ ____________.

2. We say that y **varies inversely as** the nth power of x if there is a nonzero constant k such that $y =$ ____________.

3. We say that z **varies jointly as** the nth power of x and the mth power of y if there is a nonzero constant k such that $y =$ ______________.

4. If z varies directly as the square of x and inversely as the square root of y, then $z = \frac{k}{\rule{2em}{0.4pt}}$.

Watch and Work

Watch the videos in this section, and then complete the following exercises.

1. Find the mathematical model for the verbal statement.

   T varies jointly as the cube of v and inversely as the square root of p.

2. Suppose y varies directly as the square root of x and $y = 10$ when $x = 4$. What is y when $x = 12$?

Practice

Solve the following problems in the space provided.

1. Suppose y varies inversely as the square of x and $y = 7$ when $x = 2$. What is x when $y = 9$?

2. The distance that an object falls from rest, when air resistance is negligible, varies directly as the square of the time that it falls (before it hits the ground). A stone dropped from rest travels 249 feet in the first 5 seconds. How far will it have fallen at the end of 7 seconds?

3. The volume of a gas in a container varies inversely as the pressure on the gas. If a gas has a volume of 221 cubic inches under a pressure of 5 pounds per square inch, what will be its volume if the pressure is increased to 6 pounds per square inch?

3.6 Mathematical Models

Getting Started

Topics

1. Constructing mathematical models
2. Interpolation and extrapolation

Learn

Build the foundation you need for this section by completing the following.

1. The key steps in mathematical modeling are the following:

    1. With a given question or goal in mind, identify and name the relevant ______________ quantities.
    2. Translate verbal, graphical, or other knowledge about the variables into an equation or ______________.
    3. Use the constructed equation or function to answer the given ______________.

2. To **interpolate** a value from a table or graph means to arrive at an estimate for the value of the implied function ______________ two known data points.

3. To **extrapolate** a value from a table or graph means to guess at a value of the implied function ______________ the given points.

▶ Watch and Work

Watch the videos in this section, and then complete the following exercises.

1. A person buys a car for \$15,000. The car will depreciate by 40% after five years, depreciating linearly over the course of the five years.

   (a) Construct a model for the value at time t (in years) after the initial purchase, for $0 \leq t \leq 5$.

   (b) At what point in time does the car have half its original value?

   (c) What is the value of the car after 3 years?

Practice

Solve the following problems in the space provided.

1. Construct a mathematical model as appropriate for the following situation, and then use your model to answer the accompanying questions.

   (a) Find a model for the sum of the squares of two nonnegative numbers whose sum is 20.

   (b) Expressing the sum of the squares as a function of one variable, graph the sum.

   (c) What are the largest and smallest possible sums?

2. The table below shows the distance in feet, d, that a golf ball has traveled, in feet, t seconds after it is hit with a putter.

t	0	1	2	3	4	5
d	0	15	25	29	32	33

(a) Graph the distances (either by hand or with a graphing utility).

(b) Find the linear function of best fit that models the distance traveled by the golf ball, and graph the function along with the given distances.

(c) Find the quadratic function of best fit that models the distance traveled by the golf ball, and graph the function along with the given distances.

Chapter 4

Working with Functions

Now that we have a good grasp on what functions are, we want to see what we can do with these functions. What properties do they have? Can we transform them? Can we combine them? Can we undo them? Can we stop asking questions? We'll answer all of these questions in this chapter...except maybe the last one.

4.1 Transformations of Functions

Getting Started

Now that we have a catalog of many different types of functions, we can begin to investigate how making changes to the formula of one of those functions will change the corresponding graph of the function. This section will consider a number of these transformations. You may be transformed by the experience yourself.

Topics

1. Shifting graphs vertically and horizontally
2. Reflecting graphs
3. Stretching graphs vertically and horizontally
4. Order of transformations

Learn

Build the foundation you need for this section by completing the following.

1. Let $f(x)$ be a function whose graph is known.

   (a) The graph of $g(x) = f(x-2)$ will shift the graph of f ______ units to the __________.

   (b) The graph of $g(x) = f(x) - 4$ will shift the graph of f ______ units __________.

2. Given a function $f(x)$,

   (a) the graph of the function $g(x) = -f(x)$ is the reflection of the graph of f with respect to the ______-axis.

   (b) the graph of the function $g(x) = f(-x)$ is the reflection of the graph of f with respect to the ______-axis. The graph of $g(x) = -f(x)$ will reflect the graph of f over the ______-axis.

3. Let $f(x)$ be a function and let a be a positive real number.

   (a) The graph of the function $g(x) = af(x)$ is ______________ vertically compared to the graph of f if $a > 1$.

   (b) The graph of the function $g(x) = af(x)$ is ______________ vertically compared to the graph of f if $0 < a < 1$.

4. Let $f(x)$ be a function and let a be a positive real number.

   (a) The graph of the function $g(x) = f(ax)$ is stretched horizontally compared to the graph of f by a factor of $\frac{1}{a}$ if ______________.

   (b) The graph of the function $g(x) = f(ax)$ is compressed horizontally compared to the graph of f by a factor of $\frac{1}{a}$ if ______________.

5. To construct a function g through a horizontal and vertical shift of the function f, the ______________ shift should be done before the ______________ shift.

Watch and Work

Watch the videos in this section, and then complete the following exercises.

---

1. Consider the function $f(x) = -\dfrac{1}{(x-3)^2} + 4$.

   (a) What is the more basic function that is being transformed?

   (b) Does the graph of f shift the basic function horizontally? If so, which direction and how many units?

   (c) Does the graph of f reflect the basic function over an axis? If so, which one?

   (d) Does the graph of f shift the basic function vertically? If so, which direction and how many units?

2. Sketch the graph of the function by first identifying the more basic function that has been shifted, reflected, stretched, or compressed.

$$g(x) = (x+3)^2 - 2$$

Practice

Solve the following problems in the space provided.

1. Consider the function $h(x) = \sqrt{x+3} - 5$.

   (a) Identify the more basic function that has been shifted, reflected, stretched, or compressed.

   (b) Identify the direction and number of units of the horizontal shift.

   (c) Identify the direction and number of units of the vertical shift.

2. Write a formula for the function that results when the function $g(x) = x^3$ is moved one unit to the left and reflected across the y-axis.

3. Use your knowledge about transformations to find a possible formula for the function $f(x)$ given the graph.

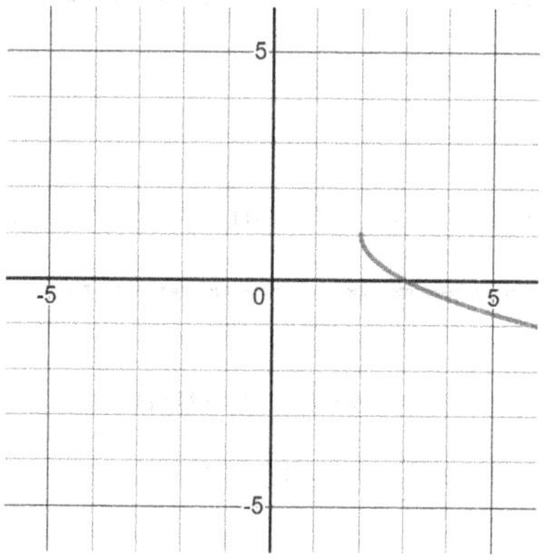

4.2 Properties of Functions

Getting Started

Being able to determine certain properties about functions can be very useful. In this section, you'll learn how to tell if functions are symmetric, where they go up and down, where their largest and smallest values are, and how they change on average. Also, you'll find out whether or not they like long walks on the beach.

Topics

1. Symmetry of functions and equations
2. Intervals of monotonicity
3. Local extrema
4. Average rate of change

Learn

Build the foundation you need for this section by completing the following.

1. The graph of a function f has **y-axis symmetry** if $f(-x) =$ ________ for all x in the domain of f. Such functions are called ________________ functions.

2. If replacing x with $-x$ and y with $-y$ in an equation results in an equivalent equation, is the equation symmetric with respect to the x-axis, y-axis, or origin?

3. We say that a function f is **decreasing on an interval** if, for any x_1 and x_2 in the interval with $x_1 < x_2$, it is the case that $f(x_1)$_______$f(x_2)$.

4. A function f has a **local minimum** at c if there is an open interval (a, b) containing c for which $f(x)$_______$f(c)$ for all x in (a, b).

5. Given a function f defined on an interval $[a, b], a \neq b$, the **average rate of change** of f over $[a, b]$ is

$$\frac{f(b) - \_\_\_\_}{\_\_\_\_\_\_\_\_}.$$

Watch and Work

Watch the videos in this section, and then complete the following exercises.

1. Determine if the function is even, odd, or neither.

$$g(x) = x^3 - x + 1$$

2. For the function, determine the location and type of the local extrema. Then determine the value(s) of the local extrema.

$$f(x) = -(x - 2)^2 - 3$$

Practice

Solve the following problems in the space provided.

1. Determine if the equation has x-axis symmetry, y-axis symmetry, origin symmetry, or none of these.

$$y = \left|\frac{x}{3}\right|$$

2. Find and identify the open intervals of monotonicity where the function is increasing, decreasing, or constant.

$$m(x) = |x - 3|$$

3. Determine the average rate of change of the function over the interval.

$$f(x) = 2x^3 - x; \;\; [0, 2]$$

4.3 Combining Functions

Getting Started

We know that we can combine the real numbers with the operations of addition, subtraction, multiplication, and division. Wouldn't it be great if we could combine functions using the same operations? As a matter of fact, we can. We'll look at those four operations and throw an extra one in (called *composition*) free of charge.

Topics

1. Combining functions arithmetically
2. Composing functions
3. Decomposing functions
4. Recursive graphics

Learn

Build the foundation you need for this section by completing the following.

1. Suppose $f(2) = 3$ and $g(2) = -4$. Find the values of the following functions.

$$\begin{aligned}(f+g)(2) &= f(2)+g(2) = \\ (f-g)(2) &= f(2)-g(2) = \\ (fg)(2) &= f(2)g(2) = \\ \left(\frac{f}{g}\right)(2) &= \frac{f(2)}{g(2)} = \end{aligned}$$

2. Let f and g be two functions. The **composition** of f and g, denoted $f \circ g$, is the function defined by $(f \circ g)(x) =$ ____________.

3. Suppose you want to decompose the function $h(x) = \sqrt[3]{2x^3 - 7}$ into the composition $f(g(x))$. What would a possible function be for $g(x)$?

4. If f is a function and $f^2(x)$ stands for $f(f(x))$, then $f^3(x) =$ ____________.

Watch and Work

Watch the videos in this section, and then complete the following exercises.

1. Let $f(x) = x^3 + 4$ and $g(x) = \sqrt{x}$.

   (a) Find $(f + g)(x)$ and its domain.

   (b) Find $\left(\frac{f}{g}\right)(x)$ and its domain.

2. Let $f(x) = \sqrt{2x - 5}+3$. Find $g(x)$ and $h(x)$ so that $f(x) = (g \circ h)(x) = g(h(x))$.

Practice

Solve the following problems in the space provided.

1. Find the formula and domain for $\frac{f}{g}$.

$$f(x) = x - 4 \text{ and } g(x) = x^{2/3}$$

2. Determine $(f \circ g)(1)$.

$$f(x) = 3 + \sqrt{x} \text{ and } g(x) = \sqrt{x+3}$$

3. Determine the formula and domain for $(f \circ g)(x)$.

$$f(x) = \sqrt{x-1} \text{ and } g(x) = \frac{x+3}{3}$$

4.4 Inverses of Functions

Getting Started

We can think of functions as a rule for sending one number to another number. For example, the function $f(x) = x^3$ sends the number $x = 2$ to the number 8 (since $2^3 = 8$). In this section, we'll develop methods for finding a function that sends the number back. This is called the *inverse* function.

Topics

1. Inverses of relations
2. Inverse functions and the horizontal line test
3. Finding the inverse of a function

Learn

Build the foundation you need for this section by completing the following.

1. Let R be a relation. The **inverse of** R, denoted R^{-1}, is the relation $R^{-1} = \{(\quad , \quad) | (a, b) \in R\}$.

2. A function f is **one-to-one** if for every pair of distinct elements x_1 and x_2 in the domain of f, we have $f(x_1)$ ______ $f(x_2)$.

3. Let f be a one-to-one function, and assume that f is defined by a formula. To find a formula for f^{-1}, perform the following steps.

Step 1: Replace $f(x)$ in the definition of f with the variable _____.

Step 2: Solve the equation for _____.

Step 3: Replace x in the remaining equation with _______ and replace each occurrence of y with x.

4. Given a function f and its inverse f^{-1}, $f\left(f^{-1}(x)\right) =$ ______ for all $x \in \text{Dom}\left(f^{-1}\right)$.

Watch and Work

Watch the videos in this section, and then complete the following exercises.

1. Let $R = \{(2,5), (-3,1), (0,1), (-7,4)\}$.

   (a) Find R^{-1}.

   (b) Find the domain and range of R^{-1}.

   (c) Is R^{-1} a function?

2. Find a formula for the inverse of the function.

$$h(x) = \sqrt[3]{3x+6}$$

Practice

Solve the following problems in the space provided.

1. Given the relation $y = 3x - 3$:

   (a) Find four points contained in the inverse of the relation.

   (b) Define the domain and range of the inverse.

2. Find a formula for the inverse of the function, if possible.

$$h(x) = 2\sqrt[5]{x} + 3$$

3. Verify that $f\left(f^{-1}(x)\right) = x$ and that $f^{-1}\left(f(x)\right) = x$.

$$f(x) = \sqrt[3]{x-3} + 2 \text{ and } f^{-1}(x) = (x-2)^3 + 3$$

Chapter 5

Polynomial and Rational Functions

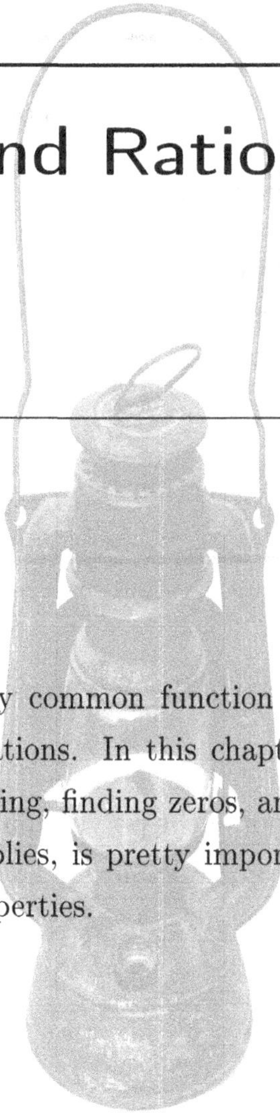

Polynomial functions are a very common function type that can occur in a wide variety of mathematical applications. In this chapter we take an in-depth look at these functions, including graphing, finding zeros, and the Fundamental Theorem of Algebra which, as its name implies, is pretty important. Finally, we'll learn about rational functions and their properties.

5.1 Polynomial Functions and Polynomial Inequalities

Getting Started

We'll begin our study of polynomial functions by looking at some of the properties of the graphs, such as behavior at infinity and factoring. We'll also take a look at polynomial inequalities, which just sound like fun.

Topics

1. Zeros of polynomial functions and solutions of polynomial equations
2. Graphing factored polynomial functions
3. Solving polynomial inequalities

Learn

Build the foundation you need for this section by completing the following.

1. The number c is said to be a **zero** of the polynomial function $p(x)$ if ______ $= 0$.

2. Fill in the blanks for the behavior of the polynomial $p(x)$ with leading term $a_n x^n$.

	Degree n is even	Degree n is odd
Leading coefficient a_n is positive	$p(x) \to$ _____ as $x \to -\infty$ $p(x) \to$ _____ as $x \to \infty$	$p(x) \to$ _____ as $x \to -\infty$ $p(x) \to$ _____ as $x \to \infty$
Leading coefficient a_n is negative	$p(x) \to$ _____ as $x \to -\infty$ $p(x) \to$ _____ as $x \to \infty$	$p(x) \to$ _____ as $x \to -\infty$ $p(x) \to$ _____ as $x \to \infty$

3. A **polynomial inequality** is any inequality that can be written in the form $p(x)$_____$0, p(x) \leq 0, p(x) > 0$, or $p(x) \geq 0$, where $p(x)$ is a polynomial function.

4. Consider the polynomial function $f(x) = (x - 4)(2x + 3)(x + 5)$.

   (a) Determine the x-coordinates of the x-intercepts of f.

   $$\{\_\_\_\_\_, \_\_\_\_\_, \_\_\_\_\_\}$$

   (b) Determine the y-intercept of f.

   $$(0, \_\_\_\_\_)$$

   (c) If the three linear factors were multiplied out, what would the leading term of the polynomial be?

Watch and Work

Watch the videos in this section, and then complete the following exercises.

---

1. Sketch the graph of the polynomial function by finding the x-intercepts and the y-intercepts and determining the behavior at infinity.

$$r(x) = (x - 2)(x + 4)(1 - x)$$

2. Solve the polynomial inequality.

$$x^2(x+3)(x-5) \leq 0$$

Practice

Solve the following problems in the space provided.

1. Verify that the given value of x solves the polynomial equation.

$$-53x^2 + 21x = 7x^3 + 24; \quad x = -8$$

2. Solve the polynomial equation by factoring or using the quadratic formula, making sure to identify all the solutions.

$$x^3 - 15x^2 + 56x = 0$$

3. Consider the polynomial $f(x) = -4x^2(x+5)(x-8)$.

   (a) Determine the degree and leading coefficient of $f(x)$.

   (b) Describe the behavior of the graph of $f(x)$ as $x \to \pm\infty$.

5.2 Polynomial Division and the Division Algorithm

Getting Started

It can be very difficult to determine the zeros of some higher-degree polynomial equations. As we have seen, however, once a polynomial is in factored form it becomes much easier to recognize the zeros of that polynomial. In this section, we will investigate how using *polynomial division* can help us find factors of a given polynomial.

Topics

1. The division algorithm and the remainder theorem
2. Polynomial long division
3. Synthetic division
4. Constructing polynomials with given zeros

Learn

Build the foundation you need for this section by completing the following.

1. Let $p(x)$ and $d(x)$ be polynomials such that $d(x) \neq 0$ and with the degree of d less than or equal to the degree of p. Then there are unique polynomials $q(x)$ and $r(x)$, called the **quotient** and the **remainder**, respectively, such that

$$p(x) = \_\_\_\_\_\_ d(x) + \_\_\_\_\_\_.$$

2. The number c is a zero of a polynomial $p(x)$ if and only if the linear polynomial __________ is a factor of p.

3. For synthetic division, if the divisor is __________, write down c, followed by the coefficients of the dividend.

4. Suppose a polynomial has zeros of $-1, -6$, and 5. What three factors must be included in the factorization of this polynomial?

Watch and Work

Watch the videos in this section, and then complete the following exercises.

1. Use polynomial long division to write the fraction in the form $q(x) + \frac{r(x)}{d(x)}$, where $d(x)$ is the denominator of the original fraction, $q(x)$ is the quotient, and $r(x)$ is the remainder.

$$\frac{5x^2 + 9x + 6}{x + 2}$$

2. Use synthetic division to determine if k is a zero of $p(x)$. If not, determine $p(k)$.

$$p(x) = 2x^3 - 8x^2 - 23x + 63; \quad k = 2$$

Practice

Solve the following problems in the space provided.

1. Use polynomial long division to write the fraction in the form $q(x) + \frac{r(x)}{d(x)}$, where $d(x)$ is the denominator of the original fraction, $q(x)$ is the quotient, and $r(x)$ is the remainder.

$$\frac{12x^4 - 18x^3 + 26x^2 + 3x + 3}{3x^2 + 2}$$

2. Use synthetic division to write the fraction in the form $q(x) + \frac{r(x)}{d(x)}$, where $d(x)$ is the denominator of the original fraction, $q(x)$ is the quotient, and $r(x)$ is the remainder.

$$\frac{5x^3 - 23ix^2 + 5x + (4 - 4i)}{x - 4i}$$

3. Construct a polynomial that is third-degree, with zeros of $-3, -1$, and 2, and passes through the point $(3, 9)$.

5.3 Locating Real Zeros of Polynomial Functions

Getting Started

If we know one of the zeros of a given polynomial, we can use division to factor that polynomial. However, sometimes it may be difficult to even locate one zero of a given polynomial. In this section, we will look at some methods designed to help us determine what numbers might possibly be zeros of a polynomial.

Topics

1. The Rational Zero Theorem
2. Descartes' Rule of Signs
3. Bounds of real zeros
4. The Intermediate Value Theorem

Learn

Build the foundation you need for this section by completing the following.

1. If $f(x) = a_n x^n + a_{n-1}x^{n-1} + \cdots + a_1 x + a_0$ is a polynomial with integer coefficients with $a_n \neq 0$, then any rational zero of f must be of the form $\frac{p}{q}$, where p is a factor of the constant term ______ and q is a factor of the leading coefficient ______.

2. How many variations in sign occur in the following polynomial?

   $h(x) = -11x^4 + 5x^3 - 7x^2 + 3x + 1$ Variations in sign : ______

3. Let $f(x)$ be a polynomial with real coefficients, a positive leading coefficient, and degree ≥ 1. Let a be a negative number and b be a positive number. Then no real zero of f is larger than b if the last row in the synthetic division of $f(x)$ by $x - b$ contains no ______________ numbers.

4. Suppose $f(x)$ is a polynomial with real coefficients and that $f(2) = -5$ and $f(3) = 2$. According to the **Intermediate Value Theorem**, there is at least one point c with _____ $< c <$ _____ and $f(c) = 0$.

Watch and Work

Watch the videos in this section, and then complete the following exercises.

1. Given the polynomial $D(x) = -2x^3 + 9x^2 + 11x - 30$:

   (a) List all of the potential rational zeros.

   (b) Use polynomial division and the quadratic formula, if necessary, to identify the actual zeros.

2. Solve the polynomial equation.

$$x^3 - 3 + 3x^2 - x = 0$$

Practice

Solve the following problems in the space provided.

1. Given the polynomial $f(x) = 2x^3 - x^2 + 8x - 4$:

   (a) List all of the possible rational zeros.

   (b) Use polynomial division and the quadratic formula, if necessary, to identify the actual zeros.

2. Does the Intermediate Value Theorem guarantee that a real zero of the polynomial exists between the indicated values? Justify your response.

$$g(x) = x^4 - 9x^3 + 64x^2 + 156x - 144; \quad \frac{11}{2} \text{ and } \frac{13}{2}$$

3. Given the polynomial $g(x) = x^3 - x^2 - 14x + 24$:

   (a) Use Descartes' Rule of Signs to determine the possible number of positive and negative zeros.

   (b) Use synthetic division to identify the best integer upper and lower bounds of the real zeros.

   (c) Using your answers to the preceding steps, polynomial division, and the quadratic formula, if necessary, list all of the zeros.

5.4 The Fundamental Theorem of Algebra

Getting Started

Well, this certainly sounds important! In this section we will combine all of the methods we learned to be able find the zeros of a multitude of polynomials. We'll show that this is never a vain attempt. In fact, the Fundamental Theorem of Algebra guarantees the existence of at least one zero for *any* polynomial of degree 1 or more. Very important, indeed.

Topics

1. The Fundamental Theorem of Algebra
2. Multiple zeros and their geometric meaning
3. Conjugate pairs of zeros
4. Summary of polynomial methods

Learn

Build the foundation you need for this section by completing the following.

1. If p is a polynomial of degree n, with $n \geq 1$, then p has at least one ________.

2. Consider the polynomial function $f(x) = x(x-2)^3(x+1)$. List the zeros of f and the multiplicity of each zero.
   Zero: _____ Multiplicity: _____ Zero: _____ Multiplicity: _____
   Zero: _____ Multiplicity: _____

3. Suppose that $3 - 5i$ is a zero of a given polynomial function. According to the **Conjugate Roots Theorem**, what complex number is also a zero?

Watch and Work

Watch the videos in this section, and then complete the following exercises.

1. Sketch the graph of the factored polynomial.

$$f(x) = x^3(x+3)(x-1)^2$$

2. Construct a third degree polynomial for which 3 is a zero of multiplicity 2, -1 is the only other zero, and the leading coefficient is 2.

Practice

Solve the following problems in the space provided.

---

1. Use all available methods to factor the polynomial completely, and then sketch the graph of the function.

$$f(x) = 2x^3 + 13x^2 + 24x + 9$$

2. Use all available methods (in particular, the Conjugate Roots Theorem, if applicable) to factor the polynomial function completely, making use of the given zero.

$$f(x) = x^4 - 10x^3 + 28x^2 - 40x + 96; \quad -2i \text{ is a zero.}$$

3. Construct a polynomial function with the following properties: fifth degree, 5 is a zero of multiplicity 3, -4 is the only other zero, leading coefficient is 3.

5.5 Rational Functions and Rational Inequalities

Getting Started

In this section, we will first deal with graphing rational functions through a series of steps including finding intercepts and *asymptotes*. Then, we'll look at how being able to graph rational functions will allow us to more easily solve rational inequalities.

Topics

1. Characteristics of rational functions
2. Vertical asymptotes
3. Horizontal and oblique asymptotes
4. Graphing rational functions
5. Solving rational inequalities

Learn

Build the foundation you need for this section by completing the following.

1. Given the rational function $f(x) = \dfrac{p(x)}{q(x)}$, the domain of f consists of all real numbers except those for which ______________.
2. The vertical line __________ is a **vertical asymptote** of a function f if $f(x)$ increases in magnitude without bound as x approaches c.
3. The horizontal line __________ is a **horizontal asymptote** of a function f if $f(x)$ approaches the value c as $x \to -\infty$ or as $x \to \infty$.
4. The first step to graph the rational function f is to factor the denominator in order to determine the ______________ of f.
5. To solve a rational inequality where the rational function $f(x) = \dfrac{p(x)}{q(x)}$ is in reduced form, first find the zeros of the ______________ $p(x)$. These values are the **zeros** of f.

Watch and Work

Watch the videos in this section, and then complete the following exercises.

1. Consider the rational function $g(x) = \dfrac{x^3 + 4}{x^2 - 9}$.

   (a) Find any vertical asymptotes

   (b) Find the oblique asymptote by using long division.

2. Solve the rational inequality.

$$\frac{3}{x^2-16} \geq \frac{x}{x^2-16}$$

Practice

Solve the following problems in the space provided.

1. Find equations for the horizontal or oblique asymptotes, if any, for the rational function.

$$f(x) = \frac{-16x^2 - 8x + 8}{4x + 7}$$

2. Consider the rational function $f(x) = \dfrac{-8}{x^2 - 25}$.

   (a) Find the vertical asymptotes, if any, of this function.

   (b) Find the horizontal or oblique asymptotes, if any, of this function.

   (c) Sketch the graph of the function, making use of your work above and additional information about intercepts and any other points that may be useful.

3. Solve the rational inequality.

$$\frac{x+3}{x+1} < \frac{x}{x+4}$$

Chapter 6

Exponential and Logarithmic Functions

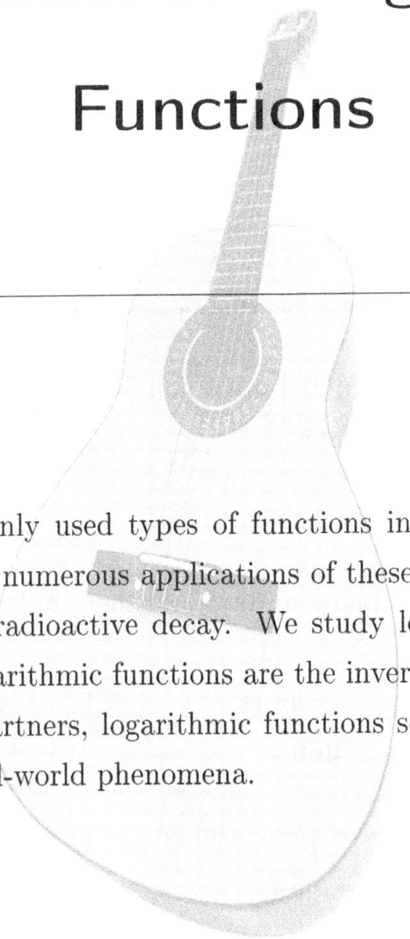

One of the most commonly used types of functions in mathematics are exponential functions. There are numerous applications of these functions from investing to population modeling to radioactive decay. We study logarithmic functions in this chapter as well, since logarithmic functions are the inverses of exponential functions. Like their exponential partners, logarithmic functions serve as an excellent tool for modeling a variety of real-world phenomena.

6.1 Exponential Functions and Their Graphs

Getting Started

In this section we take an introductory look at exponential functions. We'll investigate the properties of these functions and the basic shape of their graph. In addition, we'll try our hand at solving some basic exponential equations.

Topics

1. Characteristics of exponential functions
2. Graphing exponential functions
3. Solving elementary exponential equations

Learn

Build the foundation you need for this section by completing the following.

1. Let a be a fixed positive real number not equal to 1. The **exponential function with base a** is the function $f(x) =$ ________.

2. Recall that if $a \neq 0$, then a^0 is defined to be _____.

3. In which direction and how many units would the graph of $f(x) = 3^x$ be shifted to form the graph of $g(x) = 3^{x-2} + 4$.
   Horizontal: ____________ _____ units
   Vertical: ____________ _____ units

4. Fill in the blanks to solve the exponential equation.

$$\begin{aligned} 27^x - 3 &= 0 \\ 27^x &= \_\_\_\_ \\ \_\_\_\_^{3x} &= \_\_\_\_^{1} \\ \_\_\_\_ &= 1 \\ x &= \_\_\_\_ \end{aligned}$$

Watch and Work

Watch the videos in this section, and then complete the following exercises.

1. Sketch the graph of $g(x) = 2 - \left(\frac{3}{2}\right)^{x+3}$.

2. Solve the following exponential equation.

$$6^{4x-5} = \frac{1}{36}$$

Practice

Solve the following problems in the space provided.

1. Sketch the graph of $q(x) = \left(\frac{7}{2}\right)^{x-3}$.

2. Solve the following exponential equation.

$$\left(\frac{1}{2}\right)^{3x+2} = \left(\frac{1}{4}\right)^{2}$$

3. Solve the following exponential equation.

$$3^{-x} = 81$$

6.2 Exponential Models

Getting Started

As we mentioned, exponential functions lend themselves well to a wide variety of real-world problems. In this section, we'll take a look at just a few of these applications.

Topics

1. Models of population growth
2. Models of radioactive decay
3. Compound interest and the number e
4. Exponential regression

Learn

Build the foundation you need for this section by completing the following.

1. If the rate of growth of a population is proportional to the size of the population, the population at time t can be modeled by

$$P(t) = \underline{\qquad\qquad}.$$

2. Suppose an isotope has a half-life of 3 years. If there are initially 26 g of this isotope, how many grams will there be in 6 years?

3. Suppose \$1000 dollars is invested in a savings account at an annual interest rate of 3.5%, compounded weekly. Fill in the values for the variables in the formula $A(t) = P\left(1 + \frac{r}{n}\right)^{nt}$ that would be used to find the amount in the account after 3 years.
$P =$ ______, $r =$ _____, $n =$ _____, $t =$ _____

4. An investment of P dollars, compounded continuously at an annual interest rate of r, has a value after t years of

$$A(t) = \underline{\qquad\qquad}$$

Watch and Work

Watch the videos in this section, and then complete the following exercises.

1. The population of a certain inner-city area is estimated to be declining according to the model $P(t) = 466,000e^{-0.018t}$, where t is the number of years from the present. What does this model predict the population will be in 12 years?

2. Tessa has recently inherited \$5500, which she wants to deposit into an IRA account. She has determined that her two best bets are an account that compounds daily at an annual rate of 3.5% and an account that compounds annually at an annual rate of 3.4%. Which account would pay Tessa more interest? How much would Tessa's balance be from the higher earning account after 4 years? Round to two decimal places.

Practice

Solve the following problems in the space provided.

1. The function $C(t) = C(1+r)^t$ models the rise in the cost of a product that has a cost of C today, subject to an average yearly inflation rate of r for t years. If the average annal rate of inflation over the next 8 years is assumed to be 1.5%, what will the inflation-adjusted cost of a $32,000 motorcycle be in 8 years? Round to two decimal places.

2. Mary invests $7800 in a new savings account which earns 5.1% annual interest, compounded continuously. What will be the value of the investment after 5 years?

3. The revenue function is given by $R(x) = x \cdot p(x)$ dollars where x is the number of units sold and $p(x)$ is the unit price. If $p(x) = 41(4)^{-x/4}$, find the revenue if 16 units are sold.

6.3 Logarithmic Functions and Their Graphs

Getting Started

When we find the inverse of an exponential function, we get a logarithmic function. These two classes of functions are closely related and, since they are inverses, their basic graphs are reflections over the line $y = x$. We'll explore the graphs of logarithmic functions in more detail in this section. In addition, we'll use the properties of logarithms to evaluate logarithmic expressions and solve logarithmic equations.

Topics

1. Characteristics of logarithmic functions
2. Graphing logarithmic functions
3. Evaluating elementary logarithmic expressions
4. Solving elementary logarithmic equations
5. Common and natural logarithms

Learn

Build the foundation you need for this section by completing the following.

1. Let a be a fixed positive real number not equal to 1. The **logarithmic function with base** a is defined to be the ______________ of the exponential function with base a, and is denoted $\log_a(x)$.

2. The graph of $g(x) = \log_5(x+3) - 2$ is the graph of $f(x) = \log_5(x)$ shifted 3 units to the ____________ and _____ units down.

3. Use the fact that $\log_a(a^x) = x$ to evaluate the following logarithmic expressions.

   (a) $\log_3(27) = \log_3(3^3) =$ _____

   (b) $\log_5(\sqrt{5}) = \log_5(5^{1/2}) =$ _____

4. To solve the logarithmic equation $\log_2 x = 3$, first convert the expression to exponential form:

$$x = \_\_\_\_\_\_\_\_.$$

5. The function $\log_{10}(x)$ is called the __________ **logarithm**, and is usually written as ________.

6. The function $\log_e(x)$ is called the __________ **logarithm**, and is usually written as ________.

Watch and Work

Watch the videos in this section, and then complete the following exercises.

1. Evaluate the logarithmic expression without the use of a calculator.

$$\log\left(\sqrt[3]{10^5}\right)$$

2. Use the elementary properties of logarithms to solve the equation.

$$\log_3(5x + 2) = 3$$

Practice

Solve the following problems in the space provided.

1. Evaluate the following logarithmic expression without the use of a calculator.

$$\log_{1/8}(64)$$

2. Use the elementary properties of logarithms to solve the equation.

$$\log_{81}(x) = \frac{3}{4}$$

3. Use the elementary properties of logarithms to solve the equation.

$$6^{\log_4(x)} = 36$$

6.4 Logarithmic Properties and Models

Getting Started

As was the case with exponential functions, logarithmic functions have numerous applications to real-world problems. In this section, we'll look at some of those applications in addition to studying some further properties of logarithms.

Topics

1. Properties of logarithms
2. The change of base formula
3. Applications of logarithmic functions
4. Logarithmic regression

Learn

Build the foundation you need for this section by completing the following.

1. Fill in the blanks to complete the properties of logarithms.

   (a) $\log_a(xy) = \underline{\hspace{2cm}} + \underline{\hspace{2cm}}$

   (b) $\log_a\left(\dfrac{x}{y}\right) = \underline{\hspace{2cm}} - \underline{\hspace{2cm}}$

   (c) $\log_a(x^r) = \underline{\hspace{1cm}}\log_a(x)$

2. Rewrite the logarithmic expression $\log_7(19)$ in terms of the natural logarithm (ln) using the change of base formula.

$$\log_7(19) = \underline{\hspace{1.5cm}}$$

3. Earthquake intensity is measured on the **Richter scale** where the ranking R is given by

$$R = \log\left(\underline{\hspace{1.5cm}}\right)$$

.

4. **Logarithmic regression** is the process for finding values for a and b so that the graph of the function $f(x) =$ ______________ models the given data as well as possible.

Watch and Work

Watch the videos in this section, and then complete the following exercises.

---

1. Use the properties of logarithms to expand the expression as much as possible.

$$\ln\left(\frac{x^3\sqrt{y^6}}{e^2}\right)$$

2. Without a calculator, evaluate the expression.

$$\log_7 \frac{1}{49}$$

Practice

Solve the following problems in the space provided.

1. Use the properties of logarithms to condense the following expression as much as possible, writing the answer as a single term with a coefficient of 1.

$$\ln(x) + \ln(14y) - ln(z)$$

2. Evaluate the logarithmic expression.

$$\log_{1/2}(0.478)$$

3. If a sample of a certain solution is determined to have a $[H_3O^+]$ concentration of 5.99×10^{-5} moles/liter, what is its pH? Round off your answer to one decimal place.

6.5 Exponential and Logarithmic Equations

Getting Started

In the final section of this chapter, we focus on the interchangeability of logarithmic and exponential expressions. Somewhat ironically, to solve an exponential equation, we convert it to a logarithmic equation, and vice-versa.

Topics

1. Converting between exponential and logarithmic forms
2. Applications of exponential and logarithmic equations
3. Analysis of a stock market investment

Learn

Build the foundation you need for this section by completing the following.

1. Rewrite the exponential equation as a logarithmic equation: $2^{3x} = 15$.

2. Use the properties of logarithms to rewrite the left side of the logarithmic equation as a single logarithm.

$$\log_3(x+1) + \log_3(x-2) = \log_3(x-4)$$

3. Use properties of logarithms to fill in the blanks.

(a) $\ln(e^2) =$ ____

(b) $\log_2(x) - \log_2(x+1) =$ ______

(c) $\log_7(1) =$ ____

(d) $\log_5(5^3) =$ ____

Watch and Work

Watch the videos in this section, and then complete the following exercises.

1. Solve the exponential equation.

$$5^{3x-1} = 780$$

2. Solve the logarithmic equation.

$$\log_3(x+1) + \log_3(x) = \log_3 2$$

Practice

Solve the following problems in the space provided.

1. Solve the following exponential equation.

$$e^{3x-6} = 9^{2x/11}$$

2. Solve the following logarithmic equation.

$$\log(x-4) + \log(x+2) = \log(x-5)$$

3. How long does it take an investment to double if the investment is in a weekly-compounded savings account earning 8% per year?

Chapter 7

Trigonometric Functions

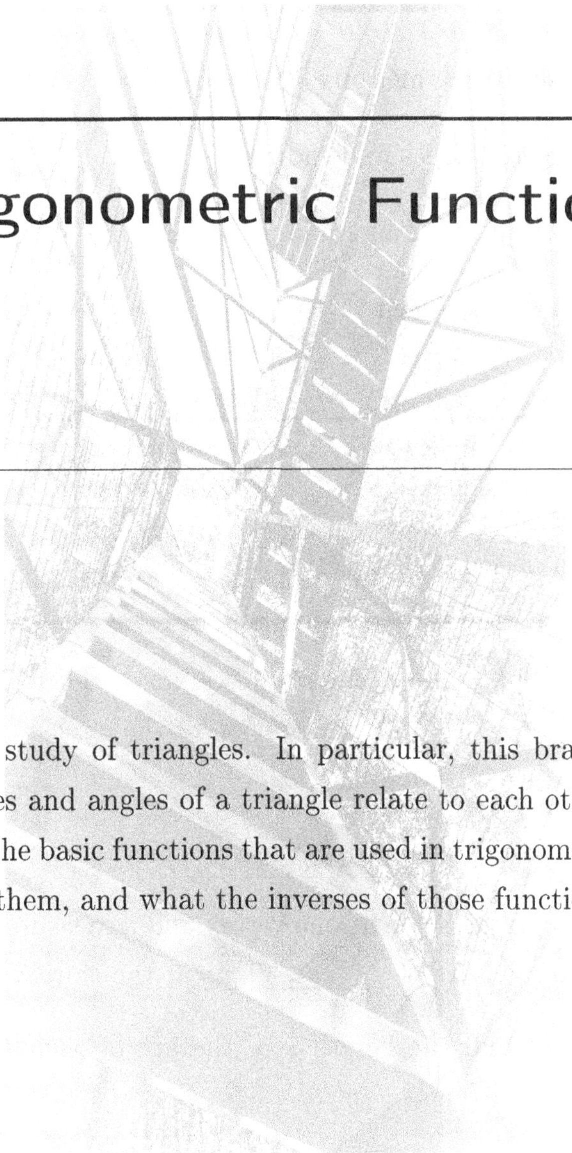

Trigonometry is the study of triangles. In particular, this branch of mathematics looks at how the sides and angles of a triangle relate to each other. In this chapter, you will learn about the basic functions that are used in trigonometry, how to evaluate them, how to graph them, and what the inverses of those functions are.

7.1 Radian and Degree Measure

Getting Started

When using the basic trigonometric functions, the values that you will put into those function are measures of angles. Angles are usually measured with one of two basic units: degrees and radians. In this section, you will learn the difference between these units and how to convert back and forth between them.

Topics

1. The unit circle and angle measure
2. Conversion between degrees and radians
3. The Pythagorean Theorem and commonly encountered angles
4. Arc length and angular speed
5. Area of a circular sector

Learn

Build the foundation you need for this section by completing the following.

1. Let θ be an angle at the center of a circle of radius 1. Then the measure of θ in **radians** is the ____________ of that portion of the circle ______________ by θ.
2. To convert degrees to radians, multiply the angle measure in degrees by ____________.
3. A triangle that has a right angle at one vertex and a measure of $\pi/3$ at another vertex, must have a measure of ____________ at the third vertex.
4. Write the formula to find the length s of the arc of a circle subtended by a central angle θ.

$$s = \_\_\_\_\_\_\_\_\_\_\_\_$$

5. The area A of a sector with a central angle of θ in a circle of radius r is

$$A = \_\_\_\_\_\_\_\_\_\_\_\_\_\_\_\_\_\_.$$

Watch and Work

Watch the videos in this section, and then complete the following exercises.

1. Frank is on a Ferris wheel that has a radius of 30 ft. If the Ferris wheel is rotating at a speed of 2 revolutions per minute, what is Frank's angular velocity?

2. Find the area of the sector of a circle of radius 10 m with a central angle of $\frac{6\pi}{5}$. Round your answer to two decimal places.

Practice

Solve the following problems in the space provided.

1. Convert the radian measure to degrees.

$$\frac{11\pi}{6}$$

2. Find the arc-length of the section of a circle of radius $r = 54$ cm subtended by the central angle $\theta = 2$.

3. Find the radian measure of the central angle θ that subtends an arc of length 29 cm on a circle of radius 8 cm.

7.2 Trigonometric Functions and Right Triangles

Getting Started

In this section, through right triangles you will be introduced to the three basic trigonometric functions (or "trig functions" for short) sine, cosine, and tangent as well as their respective reciprocal functions cosecant, secant, and cotangent. In addition, you will learn how to evaluate these functions for acute angles.

Topics

1. The trigonometric functions
2. Evaluating trigonometric functions
3. Applications of trigonometric functions

Learn

Build the foundation you need for this section by completing the following.

1. Write the formulas to evaluate the basic trig functions of an acute angle θ in a right triangle using "adj", "opp", and "hyp" for adjacent, opposite, and hypotenuse.

$$\sin\theta = \frac{\text{opp}}{\quad}, \quad \cos\theta = \frac{\quad}{\text{hyp}}, \quad \tan\theta = \underline{\qquad}$$

2. If you are given that $\cos\theta = \frac{4}{5}$, what is $\sec\theta$?

3. Use your calculator to evaluate $\csc\frac{4\pi}{9}$. Round your answer to the nearest hundredth.

4. Convert $23°37'23''$ to decimal notation.

Watch and Work

Watch the videos in this section, and then complete the following exercises.

1. For the angle θ shown in the triangle below, find $\tan\theta$ and $\csc\theta$.

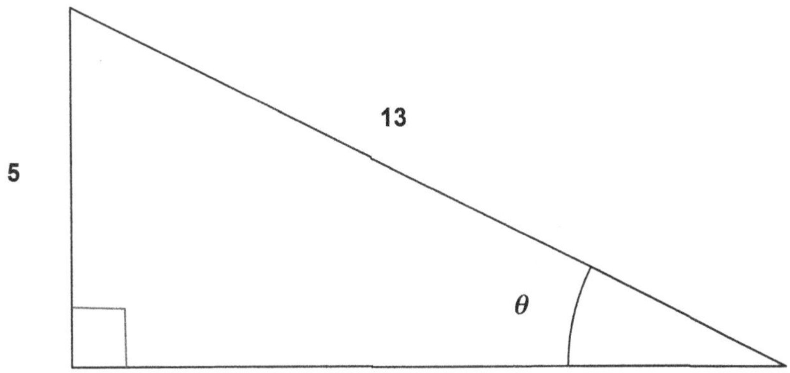

2. Evaluate the expressions, using a calculator if necessary.

$$\cos\frac{\pi}{4} \quad \text{and} \quad \sec\frac{\pi}{4}$$

Practice

Solve the following problems in the space provided.

1. Evaluate the expression, using a calculator if necessary. Round your answer to four decimal places.

$$\sec 22^\circ$$

2. Find $\tan(\theta)$ if $\cot(\theta) = 0.1$. Round your answer to four decimal places.

3. A radio tower has a 69-foot shadow cast by the sun. If the angle from the tip of the shadow to the top of the tower is 65°, what is the height of the radio tower? Round your solution to four decimal places.

7.3 Trigonometric Functions and the Unit Circle

Getting Started

In the last section, you looked at how to evaluate trig functions for an acute angle. But what if you wanted to evaluate a trig function for an angle that isn't acute? In this section, you will learn ways to do that using reference angles and the unit circle definition of the trig functions.

Topics

1. Extending the domains of the trigonometric functions
2. Evaluating trigonometric functions using reference angles
3. Relationships between trigonometric functions

Learn

Build the foundation you need for this section by completing the following.

1. Let s be a real number, and let (x, y) be the point on the unit circle associated with s. Then

$$\sin s = \_\_\_\_\_\_\_, \quad \cos s = \_\_\_\_\_\_\_, \quad \tan s = \_\_\_\_\_\_\_, x \neq 0.$$

2. Reference angles are always greater than or equal to __________ and less than or equal to __________ radians.

3. What is the reference angle associated with $\frac{7\pi}{6}$? What is $\sin \frac{7\pi}{6}$?

4. In terms of sine and cosine, how is tangent defined?

$$\tan(\theta) =$$

Watch and Work

Watch the videos in this section, and then complete the following exercises.

1. Evaluate all six trigonometric functions at the given θ.

$$\theta = \frac{5\pi}{6}$$

2. Rewrite the expression in terms of the given angle's reference angle and then evaluate the result.

$$\cos\frac{7\pi}{4}$$

Practice

Solve the following problems in the space provided.

1. Determine the reference angle associated with the given angle.

$$\theta = 315°$$

2. Given that $\sin(\theta) = 0.2588$ and $\cos(\theta) = -0.9659$, find $\cot(\theta)$. Round your answer to three decimal places, if necessary.

3. Express the following in terms of the appropriate cofunction.

$$\sin\left(\frac{9\pi}{4}\right)$$

7.4 Graphs of Sine and Cosine Functions

Getting Started

While the trigonometric functions are often thought of as functions of angles, they can also be thought of as functions of real numbers. As such, they can be graphed on the standard coordinate plane. In this section, you will investigate the graphs of the two principal trig functions, sine and cosine, and then use these functions to explore harmonic motion.

Topics

1. Graphing sine and cosine functions
2. Periodicity and symmetry
3. Amplitude, frequency, and phase shifts
4. Simple harmonic motion
5. Damped harmonic motion

Learn

Build the foundation you need for this section by completing the following.

1. A function f is said to be **periodic** if there is a positive number p such that $f(\_\_\_\_\_\_\_\_) = f(x)$ for all x in the domain of f. The smallest such number p is called the _____________ of f.

2. Complete the even/odd identities for the trig functions.

$\sin(-x) = -\sin x$	$\cos(-x) =$	$\tan(-x) =$
$\csc(-x) =$	$\sec(-x) =$	$\cot(-x) =$

3. Given constants a, b (such that $b > 0$), and c, the functions

$$f(x) = a\sin(bx - c) \text{ and } g(x) = a\cos(bx - c)$$

have **amplitude** ______, **period** _______, and a **phase shift** of _______. The left endpoint of one cycle of either function is _______ and the right endpoint is _____________.

Watch and Work

Watch the videos in this section, and then complete the following exercises.

1. Sketch the graph of $f(x) = 2\sin\left(x - \frac{\pi}{2}\right)$.

2. Sketch the graph of $g(x) = \frac{1}{2}\cos(2x + \pi)$.

Practice

Solve the following problems in the space provided.

1. Determine the amplitude, period, and phase shift of the trigonometric function.

$$y = 5 + \frac{1}{3}\sin(3x - 5)$$

Amplitude:__________ Period:__________ Phase shift:__________

2. Sketch the graph of $y = -\frac{1}{2}\cos(x + 5\pi)$.

3. Jason pulls a slinky a distance of 11 inches from its equilibrium position and then releases it. The time for one oscillation is 2 seconds. Find the amplitude and period of the slinky.

Amplitude:__________________ Period:__________________

7.5 Graphs of Other Trigonometric Functions

Getting Started

Of course sine and cosine aren't the only trig functions that can be graphed on the coordinate plane. In this section, we'll investigate the properties of the other four functions and their graphs. We'll also look at how we can graph the transformations of these functions using their periodicity, just as we did with the graphs of sine and cosine.

Topics

1. Graphing tangent and cotangent functions
2. Graphing secant and cosecant functions

Learn

Build the foundation you need for this section by completing the following.

---

1. Vertical asymptotes of the graph of tangent occur at odd multiples of __________.

2. The graph of $f(x) = 2+4\cot x$ will be the graph of $y = \cot x$ shifted ____________ by 1 unit and _______________ vertically by a factor of 4.

3. The graph of cosecant will have vertical asymptotes where ____________ $= 0$, which is every multiple of __________.

Watch and Work

Watch the videos in this section, and then complete the following exercises.

1. Sketch the graph of the function.

$$f(x) = 2\tan x - 2$$

2. Sketch the graph of the function.

$$g(x) = \frac{1}{2}\sec\left(\frac{1}{2}x - \pi\right) + 1$$

Practice

Solve the following problems in the space provided.

1. Sketch the graph of the function.

$$f(x) = -\csc x + 1$$

2. Sketch the graph of the function.

$$g(x) = 2 - \cot\left(\pi x - \frac{\pi}{2}\right) + 1$$

7.6 Inverse Trigonometric Functions

Getting Started

As is the case with other types of functions, occasionally it is useful to be able to look at the inverse of the trigonometric functions. Remember that a function has to be one-to-one in order to have an inverse, so restrictions on the domains of the trig functions must be in place to be able to consider their inverse functions.

Topics

1. The inverse trigonometric functions
2. Evaluating inverse trigonometric functions
3. Applications of inverse trigonometric functions

Learn

Build the foundation you need for this section by completing the following.

1. Given $x \in \left[-\frac{\pi}{2}, \frac{\pi}{2}\right]$, **arcsine** is defined by either of the following:

   $y = \sin x \leftrightarrow x =$ ______________ and $y = \sin x \leftrightarrow x =$ ______________

   Since the (restricted) domain of sine is $\left[\frac{-\pi}{2}, \frac{\pi}{2}\right]$ and its range is $[-1, 1]$, the domain of arcsine is ______________ and its range is ______________.

2. Fill in the table below.

Function	Notation	Domain	Range
Inverse Cosine	$\arccos y = x \leftrightarrow y =$	$[-1, 1]$	
Inverse Tangent	$\arctan y = x \leftrightarrow y =$		$\left(-\frac{\pi}{2}, \frac{\pi}{2}\right)$

3. Evaluate $\arccos\left(\cos \frac{3\pi}{4}\right)$, if possible.

Watch and Work

Watch the videos in this section, and then complete the following exercises.

1. Evaluate $\sin\left(\arccos\frac{1}{2}\right)$.

2. Evaluate $\tan(\sin^{-1}3x)$ as an algebraic function of x, assuming that $0 < x \le \frac{1}{3}$.

Practice

Solve the following problems in the space provided.

1. Solve the equation for the exact value of y without the use of a calculator.

$$y = \arctan\left(\frac{\sqrt{3}}{3}\right)$$

2. Express the following function as a purely algebraic function.

$$\cos\left(\sin^{-1}\left(\frac{2}{x}\right)\right)$$

3. Chloe is watching a rocket launch from an observation spot 2 miles away. Find the angle of elevation from Chloe to the space shuttle, which is at a height of 1.7 miles. Give your answer in degrees rounded to two decimal places.

Chapter 8

Trigonometric Identities and Equations

When working with trigonometric functions, it can be difficult to solve equations using the given functions. However, a trigonometric function can often be transformed into a simpler one. In this chapter, we will look at some of these identities and see how to use them to solve trigonometric equations.

8.1 Fundamental Trigonometric Identities

Getting Started

There are numerous trigonometric identities which can be used to write a complex trig expression as an equivalent simpler expression. Some of these are more obvious than others. In this section, you'll use basic trig identities to rewrite one expression as another and verify that two seemingly unrelated expressions are in fact equal.

Topics

1. Fundamental trigonometric identities
2. Simplifying trigonometric expressions
3. Verifying trigonometric identities
4. Trigonometric substitutions

Learn

Build the foundation you need for this section by completing the following.

1. Fill in the blanks to complete the three Pythagorean identities.

   (a) $\sin^2 x +$ ____________ $= 1$

   (b) $\tan^2 x + 1 =$ ____________

   (c) $1 +$ ____________ $= \csc^2 x$

2. Fill in the three steps used to verify trigonometric identities.

   (a) Work with ____________ ____________ at a time.

   (b) Apply ______________ ______________ as appropriate.

   (c) Rewrite in terms of ____________ and ____________ if necessary.

3. Verify the identity: $2\csc^2 x = \dfrac{1}{1-\cos x} + \dfrac{1}{1+\cos x}$.

$$\begin{aligned}
\frac{1}{1-\cos x} + \frac{1}{1+\cos x} &= \frac{1+\cos x + 1 - \cos x}{(1-\cos x)(\qquad\qquad\qquad)} \\
&= \frac{2}{1-\qquad\qquad} \\
&= \frac{2}{\qquad\qquad} \\
&= 2\csc^2 x
\end{aligned}$$

Watch and Work

Watch the videos in this section, and then complete the following exercises.

1. Use trigonometric identities to simplify the expression.

$$\frac{\tan(\beta)}{\sin(\beta)}$$

2. Verify the identity.

$$\frac{1}{\cot x} + \cot x = \sec x \csc x$$

Practice

Solve the following problems in the space provided.

1. Use trigonometric identities to simplify the expression.

$$\frac{1}{\tan^2(\theta)} + \cot(\theta) \cot\left(\frac{\pi}{2} - \theta\right)$$

2. Verify the identity.

$$\frac{1}{1-\cos^2(\beta)} = \csc^2\beta$$

3. Use the suggested substitution to write the expression as a trigonometric expression.

$$\sqrt{9x^2+36},\ \frac{x}{2} = \tan(\theta)$$

8.2 Sum and Difference Identities

Getting Started

By now, you are probably familiar with evaluating the trigonometric functions at common angles such as $\pi/3, \pi/4$, and $\pi/6$. However, you can also find the exact values of these functions at many other angles using the sum and difference identities. In this section, you'll learn how to do that as well as to use these identities to simplify other trigonometric expressions.

Topics

1. Sum and difference identities
2. Using sum and difference identities for exact evaluation
3. Using sum and difference identities for verification and simplification

Learn

Build the foundation you need for this section by completing the following.

1. Fill in the sum and difference identities for sine and cosine.

   (a) $\sin(u+v) =$ ______________ $+$ ______________

   (b) $\cos(u-v) =$ ______________ $+$ ______________

2. Use the sum formula for cosine to find the exact value of $\cos 75°$.

$$\begin{aligned}\cos 75° &= \cos(45° + \underline{\qquad}°)\\ &= \cos 45° \underline{\qquad\qquad} - \sin 45° \underline{\qquad\qquad}\\ &= \left(\frac{1}{\sqrt{2}}\right)\left(\frac{\sqrt{3}}{2}\right) - \left(\frac{1}{\quad}\right)\left(\frac{\quad}{2}\right)\\ &= \frac{\qquad}{2\sqrt{2}}\end{aligned}$$

3. Complete the Theorem of the Sum of Sines and Cosines.

$$\begin{aligned}A\sin x + B\cos x &= \sqrt{A^2+B^2}\left(\frac{A}{\sqrt{A^2+B^2}}\underline{\qquad} + \frac{B}{\sqrt{A^2+B^2}}\underline{\qquad}\right)\\ &= \sqrt{A^2+B^2}\sin(\underline{\qquad\qquad})\end{aligned}$$

where $\cos\phi = \dfrac{A}{\sqrt{A^2+B^2}}$ and $\sin\phi = \dfrac{B}{\sqrt{A^2+B^2}}$.

Watch and Work

Watch the videos in this section, and then complete the following exercises.

1. Use the sum and difference identities to determine the exact value of the expression.

$$\sin\left(\frac{7\pi}{4} - \frac{\pi}{4}\right)$$

2. Use the sum and difference identities to verify the identity.

$$\frac{\cos(x+y) - \cos(x-y)}{\sin(x+y) + \sin(x-y)} = -\tan y$$

Practice

Solve the following problems in the space provided.

1. Use the sum and difference identities to determine the exact value of the expression.

$$\sin(135^\circ)$$

2. Use the sum and difference identities to rewrite the expression as a trigonometric function of a single number, and then evaluate the result.

$$\frac{\tan(100^\circ) - \tan(45^\circ)}{1 + \tan(100^\circ)\tan(45^\circ)}$$

3. Express the following as an algebraic function of x.

$$\sin(\arctan(2x) + \arccos(2x))$$

8.3 Product-Sum Identities

Getting Started

In this section, you'll be introduced to quite a few types of trigonometric identities. Don't worry about memorizing them. Just try to be able to identify which identity to use in each situation.

Topics

1. Double-angle identities
2. Power-reducing identities
3. Half-angle identities
4. Product-to-sum and sum-to-product identities

Learn

Build the foundation you need for this section by completing the following.

1. Complete the double-angle identities.

   (a) $\sin 2u =$ ______________________

   (b) $\cos 2u =$ ______________________

   (c) $\tan 2u =$ ______________________

2. Complete the power-reducing identities.

   (a) $\sin^2 x =$ ______________________

   (b) $\cos^2 x =$ ______________________

   (c) $\tan^2 x =$ ______________________

3. Complete the half-angle identities.

   (a) $\sin \frac{x}{2} =$ ________________________

   (b) $\cos \frac{x}{2} =$ ________________________

   (c) $\tan \frac{x}{2} =$ ________________________

4. Complete the product-to-sum identities.

   (a) $\sin x \cos y =$ ________________________

   (b) $\cos x \sin y =$ ________________________

   (c) $\sin x \sin y =$ ________________________

   (d) $\cos x \cos y =$ ________________________

5. Complete the sum-to-product identities.

   (a) $\sin x + \sin y =$ ________________________

   (b) $\sin x - \sin y =$ ________________________

   (c) $\cos x + \cos y =$ ________________________

   (d) $\cos x - \cos y =$ ________________________

Watch and Work

Watch the videos in this section, and then complete the following exercises.

1. Use the information given to determine $\cos 2x, \sin 2x$,and $\tan 2x$, if possible.

2. Use the sum-to-product identities to rewrite the expression as a product.

$$\sin(3x) - \sin(x)$$

Practice

Solve the following problems in the space provided.

1. Use a power-reducing identity to rewrite the expression below in terms containing only first powers of sine and cosine.

$$\cos^4(x)\sin(x)$$

2. Determine the exact value of the expression.

$$\sin\left(\frac{11\pi}{12}\right)$$

3. Use the product-to-sum identities to rewrite the expression as a sum or a difference.

$$4\cos(4x)\cos(2x)$$

8.4 Trigonometric Equations

Getting Started

In this section, you will use the identities from the previous three sections to solve trigonometric equations. In general, the same techniques you use to solve algebraic equations will also come in handy for solving trigonometric equations. We'll also look at how to use inverse trig functions to solve trigonometric equations.

Topics

1. Solving trigonometric equations using algebraic techniques
2. Solving trigonometric equations using inverse trigonometric functions

Learn

Build the foundation you need for this section by completing the following.

---

1. One common approach to solving trigonometric equations, especially if only one trigonometric function is involved, is to ______________ the trigonometric expression on one side of the equation.

2. To solve an equation such as $\sin^2 x - \sin x = \sin x + 3$, it is helpful to realize that if $\sin x$ were replaced by u, it would simply be a ________________ equation.

3. If an equation initially appears with two or more trigonometric functions, one approach to solving it is to use an ________________ to obtain a new equation in only one function.

4. In solving real-world problems, _______________ trigonometric functions are most useful.

Watch and Work

Watch the videos in this section, and then complete the following exercises.

1. Use trigonometric identities and algebraic methods as necessary to solve the trigonometric equation.

$$1 + \cos 2x = -\sin x + 2$$

2. Use trigonometric identities and algebra as necessary to rewrite the equation to be quadratic-like, and then solve it on the interval $[0, 2\pi)$.

$$\sin^2 x - \sin x = 3 \sin x + 5$$

Practice

Solve the following problems in the space provided.

1. Use trigonometric identities and algebraic methods as necessary to solve the trigonometric equation.

$$\cos x + 2 = 2$$

2. Use trigonometric identities, algebraic methods, and inverse trigonometric functions as necessary to solve the following trigonometric equation on the interval $[0, 2\pi)$.

$$2\sin^2 x - 1 = 0$$

3. A baseball leaves a bat at an angle of θ in reference to the horizontal. The initial velocity is $v_0 = 83$ feet per second. The baseball is caught 150 feet away. What is the value of θ if the range of the baseball is given by the equation $r = \frac{1}{32}v_0^2 \sin(2\theta)$?

Chapter 9

Additional Topics in Trigonometry

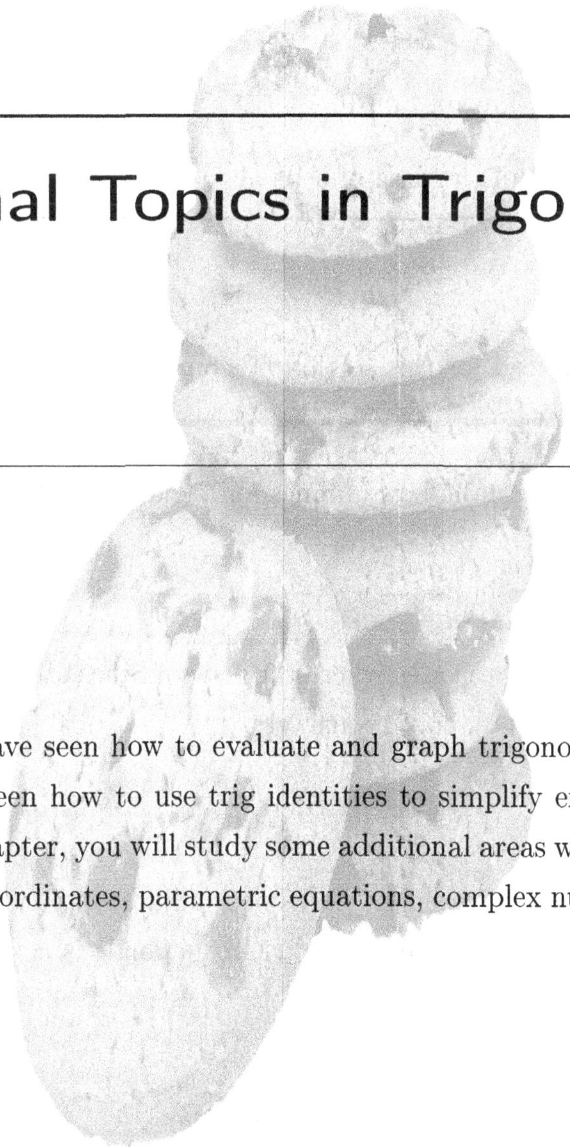

To this point, you have seen how to evaluate and graph trigonometric functions. In addition, you have seen how to use trig identities to simplify expressions and solve equations. In this chapter, you will study some additional areas where trigonometry is used such as polar coordinates, parametric equations, complex numbers, and vectors.

9.1 The Law of Sines

Getting Started

An important law that is very useful when working with triangles is the Law of Sines, which will allow us to find missing side lengths and angle measures. In this section, you'll learn this law and see how to apply it to solve problems related to triangles.

Topics

1. The Law of Sines
2. Applications of the Law of Sines
3. The Law of Sines and the area of a triangle

Learn

Build the foundation you need for this section by completing the following.

1. Fill in the blanks to write the Law of Sines for a triangle ABC with sides labeled conventionally.

$$\frac{\sin A}{\_\_\_\_} = \frac{\_\_\_\_}{b} = \frac{\_\_\_\_}{c}$$

2. Which of the following are cases where the Law of Sines allows you to solve for a given triangle? Circle all correct answers.

$$AAS, ASA, SSA, SSS, SAS$$

3. If you know two consecutive sides where one is adjacent to a given angle (SSA), how many potential triangles are there when the angle is acute?

4. Fill in the sine formula for the area of a triangle.

$$\text{Area} = \frac{1}{2}ab\sin\_\_\_\_ = \frac{1}{2}bc\sin\_\_\_\_ = \frac{1}{2}ac\sin\_\_\_\_$$

Watch and Work

Watch the videos in this section, and then complete the following exercises.

1. Solve for the remaining angle and sides of the triangles.

$$A = 45^\circ, B = 30^\circ, a = 5$$

2. Find the area of the triangle using the given information.

$$A = 125^\circ, b = 7, c = 18$$

Practice

Solve the following problems in the space provided.

1. Construct a triangle, if possible, using the following information. Round the missing lengths and angles to two decimal places.

$$A = 50^\circ, c = 6, a = 6.5$$

2. Find the area of the triangle using the given information.

$$C = 101^\circ, a = 7, b = 3$$

3. Delores is looking at a bird's nest in a tree. If she looks at it from a 65° angle and then moves back 15 feet so she can see it from a 40° angle, how high is the nest?

9.2 The Law of Cosines

Getting Started

What do we do when we are given all three sides of a triangle and need to find the angles? That's where the Law of Cosines comes in. It also works well if we know two sides and the angle between them. We'll get some practice with the Law of Cosines and its applications in this section.

Topics

1. The Law of Cosines
2. Applications of the Law of Cosines
3. Heron's Formula and the area of a triangle

Learn

Build the foundation you need for this section by completing the following.

1. Fill in the blanks to write the Law of Cosines for a triangle ABC with sides labeled conventionally.

   (a) $a^2 = b^2 + c^2 -$ ____________

   (b) ________ $= a^2 + c^2 -$ ____________

   (c) $c^2 =$ ________ $+$ ________ $- 2ab\cos C$

2. In solving triangles, it's useful to remember that the longest side of a triangle is opposite the ______________ angle.

3. Complete Heron's Formula for the area of a triangle.

   Given a triangle with sides a, b, and c, let $s = \dfrac{a+b+c}{2}$. Then

   Area = ________________________

Watch and Work

Watch the videos in this section, and then complete the following exercises.

1. Solve for the angles of the given triangle.

$$a = 5, b = 6, c = 3$$

2. Construct a triangle, if possible, using the information and the Law of Cosines.

$$A = 70^\circ, c = 10, b = 5$$

Practice

Solve the following problems in the space provided.

1. Solve for the remaining angles and side of the triangle.

$$A = 55^\circ, b = 4, c = 8$$

2. Find the area of the triangle using the given information.

$$a = 5, b = 7, c = 4$$

3. Luke creates a triangular shaped playground area in his backyard. The lengths of the sides of the areas are 5, 7, and 8 m. How large is the play area?

9.3 Polar Coordinates and Polar Equations

Getting Started

The polar coordinate system is another way of identifying unique points in the plane. Like the Cartesian coordinate system, any point in the plane can be identified by an ordered pair. In the case of the polar system, the ordered pair identifies the distance from the origin and the angle off of the polar axis. In this section, you will be introduced to the polar coordinate system as well as numerous graphs that are best described in polar coordinates.

Topics

1. The polar coordinate system
2. Converting between polar and Cartesian coordinates
3. The form of polar equations
4. Graphing polar equations

Learn

Build the foundation you need for this section by completing the following.

1. To convert from **polar to Cartesian coordinates:** Given (r, θ), x and y are defined by

   $x =$ ____________ and $y =$ ____________

2. To convert from **Cartesian to polar coordinates:** Given (x, y), r and θ are defined by

   $r^2 =$ ______________ and $\tan\theta =$ ____________, $(x \neq 0)$

3. For the polar equation $r = 2\sin\theta$, fill in the values of r for the corresponding values of θ.

θ	0	$\frac{\pi}{6}$	$\frac{\pi}{4}$	$\frac{\pi}{2}$	$\frac{2\pi}{3}$

4. An equation in r and θ is symmetric with respect to:

   (a) **The pole** if replacing r with ________ (or replacing θ with $\theta + \pi$) results in an equivalent equation;

   (b) **The polar axis** if replacing θ with ________ results in an equivalent equation;

   (c) **The vertical line** $\theta = \frac{\pi}{2}$ if replacing θ with ____________ results in an equivalent equation.

Watch and Work

Watch the videos in this section, and then complete the following exercises.

1. Convert the point from polar to Cartesian coordinates.

$$(3.5, -2\pi)$$

2. Rewrite the rectangular equation in polar form.

$$x^2 + y^2 = 49$$

Practice

Solve the following problems in the space provided.

1. Convert the point from Cartesian to polar coordinates.

$$(3, \sqrt{3})$$

2. Rewrite the polar equation in rectangular form.

$$r = 3\cos\theta$$

3. Sketch a graph of the polar equation.

$$r = 2\cos 2\theta$$

9.4 Parametric Equations

Getting Started

Parametric equations offer a new way of looking at functions and equations in the Cartesian plane and are often useful for investigating real-world applications. In this section, you will be introduced to numerous aspects of working with these important equations.

Topics

1. Applications of parametric equations
2. Graphing parametric equations by eliminating the parameter
3. Constructing parametric equations to describe a graph

Learn

Build the foundation you need for this section by completing the following.

1. Many curves in the plane are most naturally defined in terms of a variable called a ______________, a variable different from those representing the coordinates.

2. Curves described by parametric equations automatically possess an ______________, which is the direction traveled along the curve as the parameter increases.

3. The process of converting from ______________ form to ______________ form is called **eliminating the parameter**.

4. The most common method to eliminate the parameter is to solve one of the equations for the ______________ and to substitute the result for the parameter in the other ______________.

Watch and Work

Watch the videos in this section, and then complete the following exercises.

1. Construct parametric equations describing the graph of the equation.

$$y = (x - 3)^2$$

2. Construct parametric equations for the line with the given attributes.

$$\text{Slope} - 3, \text{passing through } (-2, 1)$$

Practice

Solve the following problems in the space provided.

1. Using the given values for x, construct parametric equations describing the graph of the equation.

$$y = 2x - 5, \text{ given that } x = 1 - t$$

2. Construct parametric equations for the circle with the given attributes.

$$\text{Center } (2, -3), \text{radius } 5$$

3. In a game of horseshoes, a toss is made such that the horseshoe is released 3 feet above the ground, and it leaves the pitcher's hand traveling 25 feet per second at an angle of $45°$ from the horizontal.

   (a) Construct a set of parametric equations describing the flight of the horseshoe.

   (b) How far away from the thrower will the horseshoe land?

9.5 Trigonometric Form of Complex Numbers

Getting Started

You have most likely worked with complex numbers. In this section, you will see how complex numbers can be written in a different form using trigonometry. This new form will prove useful in solving numerous types of problems involving complex numbers.

Topics

1. The complex plane
2. Complex numbers in trigonometric form
3. Multiplying and dividing complex numbers
4. Powers of complex numbers
5. Roots of complex numbers

Learn

Build the foundation you need for this section by completing the following.

1. We say the **magnitude** of a complex number $z = a + bi$, also known as the **modulus** or **absolute value** of z, is the real number $|z| =$ ________________.

2. Given $z = a + bi$, the **trigonometric form** of z is given by

   $$z = \_\_\_\_\_\_\_\_\_\_\_\_\_\_\_\_,$$

   where $|z| = \sqrt{a^2 + b^2}$ and θ satisfies $\tan\theta =$ ________. The angle θ is called the ______________ of z.

3. **Euler's Formula** states that $e^{i\theta} =$ __________ + ______________.

4. Given $z = |z|(\cos\theta + i\sin\theta)$ and a positive integer n,

   $$z^n = |z|^n \left(\cos\_\_\_\_\_\_ + i\sin\_\_\_\_\_\_\right).$$

5. Given $z = |z|(\cos\theta + i\sin\theta)$ and a positive integer n,

$$w_k = |z|^{\frac{1}{n}}\left[\cos\left(\frac{}{n}\right) + i\sin\left(\frac{}{n}\right)\right]$$

defines n distinct n^{th} roots of z for $k = 0, 1, \ldots, n-1$.

Watch and Work

Watch the videos in this section, and then complete the following exercises.

---

1. Graph and determine the magnitude of the complex number.

$$-2 + 3i$$

2. Convert the complex number to trigonometric form.

$$\sqrt{3}+i$$

Practice

Solve the following problems in the space provided.

1. Sketch z_1, z_2, z_1+z_2, and z_1z_2 on the same complex plane.

$$z_1 = 3+i, z_2 = -1+2i$$

2. Write the following complex number in standard form, $z = a + bi$.

$$4(\cos 135^\circ + i \sin 135^\circ)$$

3. Perform the operation and show the answer in both trigonometric form and standard form.

$$[2(\cos 30^\circ + i \sin 30^\circ)][3(\cos 225^\circ + i \sin 225^\circ)]$$

9.6 Vectors in the Cartesian Plane

Getting Started

Vectors are an extremely important area of mathematics and have numerous applications. In this section, you will be introduced to vectors and will learn some of the basics in describing vectors and some of their many applications.

Topics

1. Vector terminology
2. Basic vector operations
3. Component form of a vector
4. Applications of vectors

Learn

Build the foundation you need for this section by completing the following.

1. The **length** of a vector may also be referred to as its **magnitude** or ______________, and is denoted ____________.

2. Given two vectors $\mathbf{u} = \langle u_1, u_2 \rangle$ and $\mathbf{v} = \langle v_1, v_2 \rangle$ and a scalar a,

   (a) $\mathbf{u} + \mathbf{v} = \langle \_\_\_\_\_\_\_\_\_\_, \_\_\_\_\_\_\_\_\_\_ \rangle$

   (b) $a\mathbf{u} = \langle \_\_\_\_\_\_, \_\_\_\_\_\_ \rangle$

   (c) $||\mathbf{u}|| = \_\_\_\_\_\_\_\_\_\_$

3. If $\mathbf{i} = \langle 1, 0 \rangle$ and $\mathbf{j} = \langle 0, 1 \rangle$, then any vector $\mathbf{u} = \langle a, b \rangle$ can be written as $\mathbf{u} = \langle a, b \rangle = a\_\_\_\_\_ + b\_\_\_\_\_$.

4. Given the vector $\mathbf{u}$, if the magnitude $||\mathbf{u}||$ and angle with respect to the positive x-axis θ are known, then $\mathbf{u}$ can be written as

$$\mathbf{u} = \langle ||\mathbf{u}|| \cos\theta, \_\_\_\_\_\_\_\_\_\_\_\_ \rangle$$

Watch and Work

Watch the videos in this section, and then complete the following exercises.

1. With the vectors $\mathbf{u} = \langle 3, -2 \rangle$ and $\mathbf{v} = \langle 1, 2 \rangle$, find:

   (a) $-\mathbf{u}$

   (b) $2\mathbf{v} + \mathbf{u}$

   (c) $2\mathbf{v} - 3\mathbf{u}$

2. Find a unit vector pointing in the same direction as $\mathbf{u}$.

$$\mathbf{u} = \langle 3, -5 \rangle$$

Practice

Solve the following problems in the space provided.

1. Find the magnitude and direction angle of the vector.

$$\mathbf{u} = 2\mathbf{i} + 8\mathbf{j}$$

2. Find the component form of $\mathbf{v}$ given that $||\mathbf{v}|| = 6$ and $\mathbf{v}$ is in the direction of $\mathbf{i} + 5\mathbf{j}$.

3. A football was thrown into the air at a speed of 58 miles per hour at an angle of 25° from the horizontal. Express this velocity in vector form. Round your answer to four decimals.

9.7 The Dot Product

Getting Started

There are many operations that can be performed with vectors, and the dot product is one of the most important. In this section, you will learn how to find the dot product of two vectors as well as an important real-world application of the dot product.

Topics

1. Properties of the dot product
2. Projections of vectors
3. Applications of the dot product

Learn

Build the foundation you need for this section by completing the following.

---

1. Given two vectors $\mathbf{u} = \langle u_1, u_2 \rangle$ and $\mathbf{v} = \langle v_1, v_2 \rangle$, the **dot product $\mathbf{u} \cdot \mathbf{v}$** of the two vectors is the scalar defined by

$$\mathbf{u} \cdot \mathbf{v} = \underline{\qquad\qquad} + \underline{\qquad\qquad}.$$

2. Let two nonzero vectors $\mathbf{u}$ and $\mathbf{v}$ be depicted so that their initial points coincide, and let θ be the smaller of the two angles formed by $\mathbf{u}$ and $\mathbf{v}$. Then

$$\mathbf{u} \cdot \mathbf{v} = ||\mathbf{u}||\,||\mathbf{v}||\underline{\qquad\qquad}.$$

3. Two nonzero vectors $\mathbf{u}$ and $\mathbf{v}$ are __________________ if $\mathbf{u} \cdot \mathbf{v} = 0$.

4. Let $\mathbf{u}$ and $\mathbf{v}$ be nonzero vectors. The **projection of u onto v** is the vector

$$\mathbf{proj_v u} = \left(\frac{\qquad\qquad}{||\mathbf{v}||}\right)\mathbf{v}$$

.

Watch and Work

Watch the videos in this section, and then complete the following exercises.

1. Calculate the dot product.

$$\langle 1, 9\rangle \cdot \langle 7, 5\rangle$$

2. Find the projection of $\mathbf{u}$ onto $\mathbf{v}$.

$$\mathbf{u} = \langle -7, 1\rangle, \mathbf{v} = \langle 7, 4\rangle$$

Practice

Solve the following problems in the space provided.

1. Find the angle between the given vectors.

$$\mathbf{u} = \cos\left(\frac{5\pi}{4}\right)\mathbf{i} + \sin\left(\frac{5\pi}{4}\right)\mathbf{j}, \quad \mathbf{v} = \cos\left(\frac{3\pi}{4}\right)\mathbf{i} + \sin\left(\frac{3\pi}{4}\right)\mathbf{j}$$

2. Find $\mathbf{u} \cdot \mathbf{v}$ where θ is the angle between $\mathbf{u}$ and $\mathbf{v}$.

$$||u|| = 11, ||v|| = 4, \theta = \frac{\pi}{3}$$

3. A car with a gross weight of 4000 pounds is parked on a 13° slope. What force is required to prevent the car from rolling down the hill?

9.8 Hyperbolic Functions

Getting Started

Aside from their annoying habit of exaggerating everything, hyperbolic functions play a useful role in mathematics and have many applications. They are so named since they relate to a hyperbola in the same way that the trigonometric functions relate to a circle. In this section, we will define the hyperbolic functions, look at some identities, and investigate the inverses of those functions.

Topics

1. The hyperbolic functions
2. Hyperbolic identities
3. The inverse hyperbolic functions

Learn

Build the foundation you need for this section by completing the following.

1. Complete the formulas for the hyperbolic functions.

   (a) $\sinh x = \dfrac{\quad}{2}$

   (b) $\cosh x = \dfrac{\quad}{2}$

   (c) $\tanh x = \dfrac{e^x - e^{-x}}{\quad}$

2. Complete the hyperbolic identities.

   (a) $\cosh^2 x$_____$\sinh^2 x = 1$

   (b) $\cosh 2x = \cosh^2 x +$ ____________

   (c) $\sinh(x + y) =$ ____________ $\cosh y +$ ____________ $\sinh y$

Watch and Work

Watch the videos in this section, and then complete the following exercises.

1. Evaluate the expression.

$$\cosh 2$$

2. Verify the identity.

$$e^{2x} = \cosh 2x + \sinh 2x$$

Practice

Solve the following problems in the space provided.

1. Evaluate the expression.

$$\cosh(\ln 2)$$

2. Evaluate the expression.

$$\operatorname{sech}^{-1} 1$$

3. Evaluate the expression, rounding your answer to two decimal places.

$$\sinh^{-1}(-2)$$

Chapter 10

Conic Sections

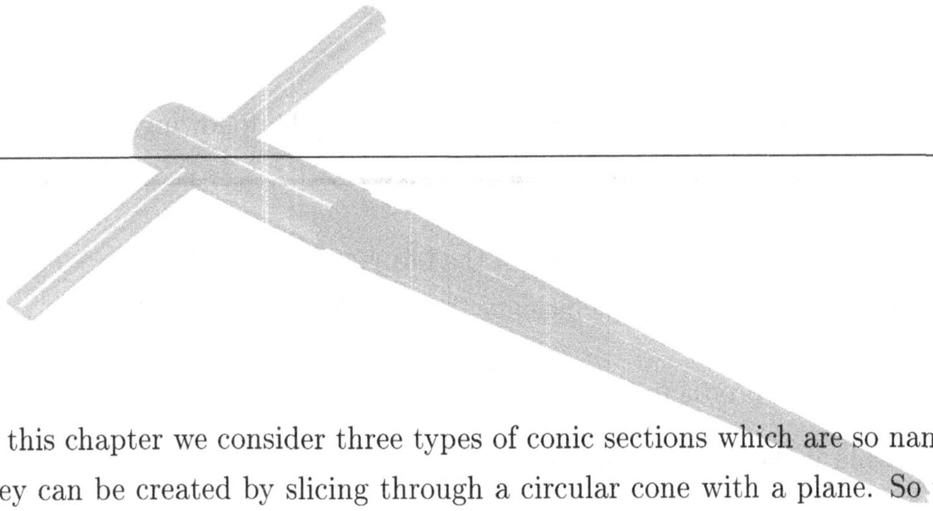

In this chapter we consider three types of conic sections which are so named because they can be created by slicing through a circular cone with a plane. So unlike some cones, we only have 3 flavors: the ellipse, the parabola, and the hyperbola. That's a lot fewer than 31.

10.1 Ellipses

Getting Started

We begin with an overview of conic sections and the study of ellipses. As we shall see, there are quite a few things that can be modeled with conic sections and ellipses are no exception. We'll look at some of those applications after we get a working understanding of the standard form of an ellipse.

Topics

1. Overview of conic sections
2. The standard form of the equation of an ellipse
3. Applications of ellipses

Learn

Build the foundation you need for this section by completing the following.

1. A conic section described by an equation of the form $Ax^2+Cy^2+Dx+Ey+F=0$, where at least one of the two coefficients A and C is not equal to 0, is:
   (a) an **ellipse** if the product AC is ______________;
   (b) a **parabola** if the product AC is _____;
   (c) a **hyperbola** if the product AC is ______________.
2. Consider the ellipse given by the equation $\frac{(x-2)^2}{4}+\frac{(y+3)^2}{16}=1$.
   (a) The center of the ellipse is at the point (_____,_____).
   (b) The major axis is vertical and has length _____.
   (c) The minor axis is horizontal and has length _____.
3. Kepler's First Law of Planetary Motion states that the planets orbit the sun in elliptical paths, with the sun at one ______________ of each orbit.

Watch and Work

Watch the videos in this section, and then complete the following exercises.

1. Find the center, foci, and vertices of the ellipse that the equation describes.

$$x^2 + \frac{(y-3)^2}{49} = 1$$

2. Sketch the graph of the ellipse and determine the coordinates of the foci.

$$16x^2 + 64y^2 + 160x + 512y + 400 = 0$$

Practice

Solve the following problems in the space provided.

1. Find the equation, in standard form, for the ellipse which is centered at $(-4, -5)$, has major axis of length 18 oriented vertically, and minor axis of length 6.

2. Find the eccentricity and the lengths of the minor and major axes of the ellipse.

$$\frac{x^2}{49} + \frac{y^2}{25} = 1$$

3. The planet Uranus moves in an elliptical orbit with the sun at one focus. Given that Uranus' closest approach to the sun is approximately 2741.3 million kilometers and that the eccentricity of Uranus' orbit is approximately 0.046, estimate this planet's maximum distance from the sun.

10.2 Parabolas

Getting Started

The next conic section we will look at is the parabola. While we have studied parabolas when graphing quadratic functions, in this section we take a more in-depth look at the properties and equations of parabolas, including ones that open left or right.

Topics

1. The standard form of the equation of a parabola (as a conic section)
2. Applications of parabola

Learn

Build the foundation you need for this section by completing the following.

1. Let p be a nonzero real constant. The **standard form of the equation for the parabola** with vertex at (h, k) is: $(x-h)^2 = 4p(y-k)$ if the parabola is __________ oriented. In this case, the focus is at (_____, __________) and the equation of the directrix is $y =$ __________);

2. Given the equation of the parabola $(x-2)^2 = 12(y+3)$, determine the following.

   (a) The vertex of the parabola: (_____, -3)

   (b) The orientation of the parabola (vertical or horizontal): ____________

   (c) The value of p (*note:* $4p = 12$): __________

3. The height $h(t)$ of an object thrown under the influence of gravity with initial velocity v_0 and initial height h_0 is

   $$h(t) = \underline{\hspace{6cm}}$$

   where g is the acceleration due to gravity.

Watch and Work

Watch the videos in this section, and then complete the following exercises.

1. Graph the parabola and determine the focus and directrix.

$$x^2 - 6x + 4y = -9$$

2. Find the equation, in standard form, for the parabola with the given properties.

Vertex at $(12, -13)$, focus at $(12, 2)$.

Practice

Solve the following problems in the space provided.

1. Graph the parabola and determine the focus and directrix.

$$y^2 - 8y = 8x + 8$$

2. Find the equation, in standard form, for the parabola with the given properties.

Symmetric with respect to the line $y = -3$, directrix is the line $x = -1$, and $p = -1$.

3. A spotlight is made by placing a strong light bulb inside a reflective paraboloid formed by rotating the parabola $x^2 = 5y$ around its axis of symmetry (assume that x and y are in units of inches). In order to have the brightest, most concentrated light beam, how far from the vertex should the bulb be placed?

10.3 Hyperbolas

Getting Started

The final conic section to consider in this chapter is the *hyperbola*. Hyperbolas share some similarities with other conic sections, in particular with parabolas, but there are some key differences that we will explore in this section.

Topics

1. The standard form of the equation of a hyperbola
2. Applications of hyperbolas

Learn

Build the foundation you need for this section by completing the following.

1. For the points on a hyperbola, the magnitude of the ______________ of the distances to two foci is a fixed constant.

2. The standard form of the equation for the hyperbola with center at (h, k) is:

   (a) $\dfrac{(x-h)^2}{a^2} - \dfrac{(y-k)^2}{b^2} = 1$ if the foci are aligned ______________;

   (b) $\dfrac{(y-k)^2}{a^2} - \dfrac{(x-h)^2}{b^2} = 1$ if the foci are aligned ______________.

3. Consider the hyperbola given by the equation $\dfrac{(x-3)^2}{4} - \dfrac{(y+1)^2}{16} = 1$.

   (a) The vertices of the hyperbola are at the point (_____, _____) and (_____, _____).

   (b) The foci of the parabola are aligned ______________.

   (c) The foci of the hyperbola are at the points (_____, -1) and (_____, -1).

   (d) The slopes of the asymptotes of the hyperbola are $\pm$_____.

Watch and Work

Watch the videos in this section, and then complete the following exercises.

1. Sketch the graph of the hyperbola using asymptotes as guides. Determine the coordinates of the foci.

$$\frac{(x+3)^2}{49} - \frac{(y+4)^2}{16} = 1$$

2. Find the equation, in standard form, of the hyperbola with the given properties.

   Foci at $(-1,3)$ and $(-1,13)$ and the asymptotes of $y-8=\pm\frac{3}{4}(x+1)$.

Practice

Solve the following problems in the space provided.

1. Find the center, foci, and vertices of the hyperbola that the equation describes.

$$\frac{(y-4)^2}{16} - (x+8)^2 = 1$$

2. Find the equation, in standard form, of the hyperbola with the given properties.

   Foci at $(4, -2)$ and $(12, -2)$ and the vertices at $(5, -2)$ and $(11, -2)$.

3. Find the equation, in standard form, for the hyperbola shown in the graph.

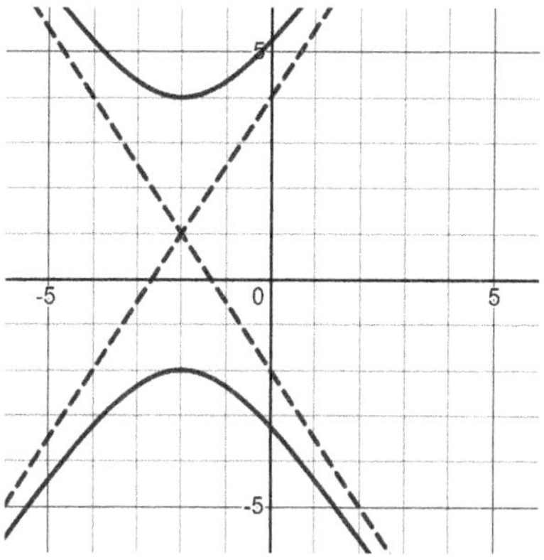

10.4 Rotation of Conic Sections

Getting Started

To this point, you have worked with conic sections that are oriented horizontally or vertically. This raises the question of how to write the equations of conic sections that are rotated an angle of θ off of the horizontal. In this section, you will learn how to handle these rotated conic sections.

Topics

1. Rotation relations
2. Rotation invariants

Learn

Build the foundation you need for this section by completing the following.

1. Given a rectangular coordinate system with axes x' and y' rotated by angle θ with respect to axes x and y, the two sets of coordinates (x', y') and (x, y) for the same point are related by:

   $x = x'$__________ $-$ __________ $x' =$ __________ $+\, y$__________

   $y =$ __________ $+\, y'$__________ $y' = -x$__________ $+$ __________

2. The graph of the equation $Ax^2+Bxy+Cy^2+Dx+Ey+F=0$ in the xy-plane is the same as the graph of the equation $A'x'^2+C'y'^2+D'x'+E'y'+F'=0$ in the $x'y'$-plane, where the angle of rotation θ between the two coordinate systems satisfies

   $$\cot 2\theta = \frac{\rule{3cm}{0.4pt}}{B}.$$

3. Assuming the graph of $Ax^2+Bxy+Cy^2+Dx+Ey+F=0$ is a nondegenerate conic section, it is classified by its discriminant as follows:

   1. ________________ if $B^2-4AC<0$
   2. ________________ if $B^2-4AC=0$
   3. ________________ if $B^2-4AC>0$

Watch and Work

Watch the videos in this section, and then complete the following exercises.

1. Use the discriminant to determine whether the equation of the given conic section represents an ellipse, a parabola, or a hyperbola.

$$4x^2 - 2xy + 6y^2 + 9x - 6y - 10 = 0$$

2. Determine the angle θ that will allow you to convert the equation and eliminate the xy-term.

$$4x^2 - 2xy + 6y^2 + 9x - 6y - 10 = 0$$

Practice

Solve the following problems in the space provided.

1. Find the $x'y'$-coordinates for the given point and the given angle of rotation θ.

$$(-13, -8\sqrt{2}), \quad \theta = \frac{\pi}{3}$$

2. Use the discriminant to classify the conic section. Then determine the angle θ that you would use to convert the equation and eliminate the xy-term.

$$-3\sqrt{3}x^2 - 21xy + 4y^2 - 6x - 10y + 2 = 0$$

3. Sketch the graph of the conic section.

$$3x^2 - 4xy + y^2 = 12$$

10.5 Polar Equations of Conic Sections

Getting Started

So far in this chapter, you have looked at conic sections as described by equations using rectangular coordinates. It turns out that polar coordinates can also be used to write the equations of conic sections. That will be the focus of this section.

Topics

1. The focus and directrix of a conic section
2. Graphing polar equations of conic sections

Learn

Build the foundation you need for this section by completing the following.

1. Let $D(P, F)$ denote the distance between a variable point P and the fixed ____________ F, and let $D(P, L)$ denote the shortest distance between P and the fixed ____________ L. A conic section consists of all points P in the plane that satisfy the equation

$$\frac{D(P, F)}{} = e,$$

where e is a fixed positive constant.

2. The conic is:

   (a) an ____________ if $0 < e < 1$,

   (b) a ____________ if $e = 1$, and

   (c) a ____________ if $e > 1$.

3. Fill in the table describing conic sections with focus at the origin and either a horizontal or vertical directrix.

Conic Section Equation	**Directrix**
$r = \dfrac{ed}{\rule{2cm}{0.4pt}}$	Vertical Directrix $x = d$
$r = \dfrac{ed}{1 - e\cos\theta}$	Vertical Directrix ________
$r = \dfrac{ed}{1 + e\sin\theta}$	____________ Directrix $y = d$
$r = \dfrac{ed}{\rule{2cm}{0.4pt}}$	Horizontal Directrix $y = -d$

Watch and Work

Watch the videos in this section, and then complete the following exercises.

1. Identify the conic section and find the equation for its directrix.

$$r = \frac{8}{7 - 6\cos\theta}$$

2. Sketch the graph of the conic section.

$$r = \frac{8}{7 - 6\cos\theta}$$

Practice

Solve the following problems in the space provided.

1. Construct a polar equation for the conic section described.
   Conic: hyperbola; Eccentricity: $e = 8$; Directrix: $y = -11$

2. Sketch the graph of the conic section.

$$r = \frac{2}{4 + 4\cos\theta}$$

3. Sketch the graph of the conic section.

$$r = \frac{6}{7 + 2\sin\theta}$$

Chapter 11

Systems of Equations and Inequalities

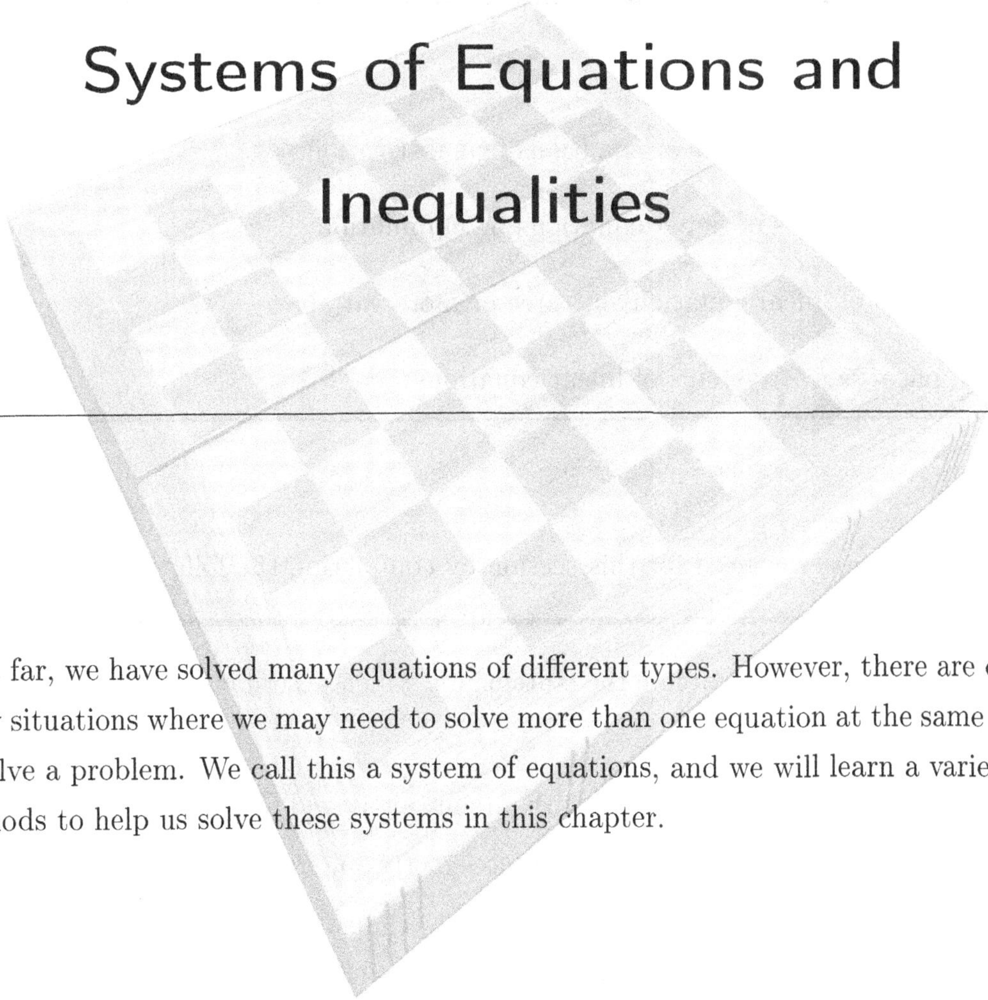

Thus far, we have solved many equations of different types. However, there are quite a few situations where we may need to solve more than one equation at the same time to solve a problem. We call this a system of equations, and we will learn a variety of methods to help us solve these systems in this chapter.

11.1 Solving Systems of Linear Equations by Substitution and Elimination

Getting Started

One of the most basic types of systems of equations is a linear system. There are a few ways to solve a system of linear equations, but two of the most common are by using substitution and elimination. We'll explore both of these methods in this section as well as look at some possible applications of linear systems.

Topics

1. Systems of linear equations
2. Solving systems of linear equations by substitution
3. Solving systems of linear equations by elimination
4. Systems of linear equations in three or more variables
5. Applications of systems of linear equations

Learn

Build the foundation you need for this section by completing the following.

1. Fill in the blanks to complete the statements, which describe the three varieties of linear equations.

   (a) A linear system of equations with no solution is said to be ______________.

   (b) A linear system of equations that has exactly one solution is called ______________

   (c) A linear system of equations with more than one solution must have an infinite number of solutions, and is said to be ______________.

2. Consider the system $\begin{cases} 2x + 4y &= 8 \\ 3x - y &= -9 \end{cases}$. If solving this system by substitution, it would be easiest to solve equation (2) for y.

   (a) Solve equation (2) for y: $y =$ ____________

(b) Now substitute your answer from the previous step in for y in Equation (1).

$2x + 4(\_\_\_\_\_\_\_\_\_\_) = 8$

(c) Now solve the remaining equation from the previous step for x.

$x = \_\_\_\_$

(d) Finally, use that value for x to find the corresponding y value by substituting into the equation you found in Step 1.

$y = 3(\_\_\_\_) + 9 = \_\_\_\_$

3. Consider the system $\begin{cases} 3x - 2y &= -1 \\ 5x + 3y &= 11 \end{cases}$. Suppose you wanted to solve this system by elimination. In order to eliminate y, you could multiply Equation (1) by 3 and Equation (2) by 2. Perform these multiplications to get a new system.

$$\begin{cases} 3x - 2y &= -1 \\ 5x + 3y &= 11 \end{cases} \longrightarrow \begin{cases} \_\_\_\_ - 6y &= \_\_\_\_ \\ \_\_\_\_ + 6y &= \_\_\_\_ \end{cases}$$

4. The sum of three integers is 41. The sum of the first and third integers exceeds the second by 11. The first integer is 10 less than the second. Set up a system of three linear equations that would be used to solve the system. Use the variables x, y, and z for the first, second, and third integers, respectively.

$$\begin{cases} x + y + z &= \_\_\_\_ \\ x + z &= \_\_\_\_ + 11 \\ x &= \_\_\_\_\_\_ \end{cases}$$

Verify that $x = 5, y = 15$, and $z = 21$ is the solution to this system by making sure those values make all three equations true.

Watch and Work

Watch the videos in this section, and then complete the following exercises.

1. Use the method of substitution to solve the system of equations. If the system is dependent, express the solution set in terms of one of the variables.

$$\begin{cases} -2x &= 14 \\ -4x + 5y &= 13 \end{cases}$$

2. Use the method of elimination to solve the system of equations. If the system is dependent, express the solution set in terms of one of the variables.

$$\begin{cases} 4x - 2y &= -10 \\ -3x + y &= 6 \end{cases}$$

Practice

Solve the following problems in the space provided.

1. Use any convenient method to solve the system of equations. If the system is dependent, express the solution set in terms of one of the variables.

$$\begin{cases} 2x - 4y &= 18 \\ -10x + 20y &= -91 \end{cases}$$

2. Use any convenient method to solve the system of equations. If the system is dependent, express the solution set in terms of one or more of the variables, as appropriate.

$$\begin{cases} x + 3y &= -11 \\ y - 6z &= 3 \\ -2x - 5y + 2z &= 21 \end{cases}$$

3. Use a system of equations to solve the problem.
   How many ounces of a 17% alcohol solution and a 34% alcohol solution must be combined to obtain 51 ounces of a 32% solution?

11.2 Matrix Notation and Gauss-Jordan Elimination

Getting Started

In the last section, we used the elimination method to solve systems of linear equations. In this section, we will codify that method using matrix notation and Gaussian elimination. This is often a preferable way to solve larger systems of equations since it can be done systematically.

Topics

1. Systems of linear equations, matrices, and augmented matrices
2. Gaussian elimination and row echelon form
3. Gauss-Jordan elimination and reduced row echelon form

Learn

Build the foundation you need for this section by completing the following.

1. Given the matrix $A = \begin{bmatrix} 3 & -2 & 11 \\ 0 & 6 & -5 \\ 4 & 15 & -21 \\ 31 & -16 & 1 \end{bmatrix}$, determine the order of A is ______ $\times$ ______ and the value of a_{32} is ______

2. Convert the system of linear equations to an augmented matrix.

$$\begin{cases} 4x - 6y = 18 \\ -2x + 5y = 7 \end{cases} \rightarrow \left[\begin{array}{cc|c} -2 & \_\_\_ & \_\_\_ \\ \_\_\_ & \_\_\_ & 7 \end{array}\right]$$

3. A matrix is in **row echelon form** if:

   1. The first non-zero entry in each row is _____. We call this a **leading 1**.
   2. Every entry below a leading 1 is _____, and each leading 1 appears farther to the _________ than the leading 1's in the rows above it.
   3. All rows consisting entirely of _____ (if there are any) appear at the bottom.

4. Perform the indicated row operation to the augmented matrix.

$$\left[\begin{array}{ccc|c} 1 & -2 & 5 & 3 \\ 2 & 4 & -4 & 0 \\ 3 & -1 & 6 & -5 \end{array}\right] \xrightarrow{-3R_1+R_3} \left[\begin{array}{ccc|c} 1 & 2 & 5 & 3 \\ 2 & 4 & -4 & 0 \\ & & & \end{array}\right]$$

5. A matrix is said to be in **reduced row echelon form** if:

   1. It is in _____ _____________ __________.

   2. Each entry __________ a leading 1 is also 0.

Watch and Work

Watch the videos in this section, and then complete the following exercises.

---

1. Use Gaussian elimination and back-substitution to solve the system of equations.

$$\begin{cases} x - 5y &= -7 \\ 4x + 3y &= 18 \end{cases}$$

2. Use Gauss-Jordan elimination to solve the system of equations.

$$\begin{cases} x + 4y &= 14 \\ y - 3z &= 1 \\ -2x - 3y + 5z &= -3 \end{cases}$$

Practice

Solve the following problems in the space provided.

1. Use Gaussian elimination and back-substitution to solve the system of equations.

$$\begin{cases} 3x - 4y &= 16 \\ -6x + 8y &= -34 \end{cases}$$

2. Use Gauss-Jordan elimination to solve the system of equations.

$$\begin{cases} 2x + 7y &= -12 \\ 6x + 5y &= 2 \end{cases}$$

3. Use Gauss-Jordan elimination to solve the system of equations.

$$\begin{cases} -3x + 2y + 2z &= -14 \\ x + 5y - 4z &= 27 \\ 4x + 3y - 6z &= 41 \end{cases}$$

11.3 Determinants and Cramer's Rule

Getting Started

In this section, we investigate a method for solving linear systems of equations called Cramer's Rule. In order to use this method, we'll first need to learn how to find the determinant of a matrix. Determinants play a very important role in the area of linear algebra, and we'll see how they can help us solve systems of equations.

Topics

1. Evaluating determinants
2. Solving systems of linear equations using Cramer's Rule

Learn

Build the foundation you need for this section by completing the following.

1. Find the determinant of the 2×2 matrix $A = \begin{bmatrix} 3 & 7 \\ -2 & 4 \end{bmatrix}$.

$$|A| = (3)(\_\_\_\_) - (-2)(\_\_\_\_) = \_\_\_\_ - (-14) = \_\_\_\_$$

2. Let A be an $n \times n$ matrix and let i and j be two fixed numbers each between 1 and n.

   (a) The **minor** of the element a_{ij} is the determinant of the $(n-1) \times (n-1)$ matrix formed from A by deleting its ____ row and ____ column.

   (b) The **cofactor** of the element a_{ij} is ______ times the minor of a_{ij}.

3. In evaluating the determinant of the following matrix, along which row or column should you expand to make the calculations the easiest? Why?

$$A = \begin{bmatrix} -2 & 5 & 0 \\ 3 & 4 & 0 \\ 10 & -7 & 2 \end{bmatrix}$$

4. Consider the system of equations $\begin{cases} 2x + 3y &= 8 \\ -x + 4y &= 7 \end{cases}$. Find the following.

(a) $D = \begin{vmatrix} 2 & 3 \\ -1 & 4 \end{vmatrix} =$

(b) $D_x = \begin{vmatrix} & 3 \\ & 4 \end{vmatrix} =$

(c) $D_y = \begin{vmatrix} 2 & \\ -1 & \end{vmatrix} =$

(d) $x = \dfrac{D_x}{D} = \underline{\hspace{2em}} =$

(e) $y = \dfrac{D_y}{D} = \underline{\hspace{2em}} =$

Watch and Work

Watch the videos in this section, and then complete the following exercises.

---

1. Evaluate the determinant of the matrix.

$$A = \begin{bmatrix} -7 & 19 \\ 13 & 21 \end{bmatrix}$$

2. Use Cramer's rule to solve the system.

$$\begin{cases} x + 22y & = & -140 \\ y + 15z & = & -66 \\ 7x + 16z & = & -120 \end{cases}$$

Practice

Solve the following problems in the space provided.

1. Determine the cofactor of a_{13} in the matrix.

$$A = \begin{bmatrix} 5 & -8 & 5 \\ 3 & 5 & 6 \\ 1 & 1 & -6 \end{bmatrix}$$

2. Evaluate the determinant of the matrix.

$$A = \begin{bmatrix} 6 & 0 & 5 \\ -9 & 5 & 0 \\ 7 & -5 & 5 \end{bmatrix}$$

3. Use Cramer's rule to solve the system.

$$\begin{cases} 7x - 3y &= -19 \\ 8x + 2y &= -38 \end{cases}$$

11.4 Basic Matrix Operations

Getting Started

Now that we have worked with matrices to solve linear systems of equations, it's time that we take a more in-depth look at these objects. As is the case with numbers, we can perform many operations with matrices including addition, subtraction, and multiplication. These operations form an *algebra* of matrices. We will investigate these operations, as well as some of their restrictions, in this section.

Topics

1. Matrix addition
2. Scalar multiplication
3. Matrix multiplication
4. Transition matrices

Learn

Build the foundation you need for this section by completing the following.

1. Two matrices A and B can be added to form the new matrix $A + B$ only if A and B are of the same ______________.

2. Note that the following matrices A and B are both of order 2×3, so they can be added together. Find $A + B$.

$$A = \begin{bmatrix} -5 & 7 & -10 \\ 3 & 8 & 4 \end{bmatrix}, \quad B = \begin{bmatrix} 15 & -9 & 1 \\ 0 & 14 & -6 \end{bmatrix}, \quad A+B = \begin{bmatrix} 10 & \_\_\_ & \_\_\_ \\ \_\_\_ & 22 & \_\_\_ \end{bmatrix}$$

3. Find the matrix $-2A$ for $A = \begin{bmatrix} 4 & -8 \\ 0 & 12 \\ -6 & 9 \end{bmatrix}$.

$$-2A = \begin{bmatrix} -8 & \_\_\_ \\ \_\_\_ & -24 \\ \_\_\_ & -18 \end{bmatrix}$$

4. Two matrices A and B can be multiplied together, resulting in a new matrix denoted AB, only if the number of ______________ of A (the matrix on the left) is the same as the number of ______________ of B (the matrix on the right).

5. Consider the two matrices below. Note that A is of order 2×3 and B is of order 3×2 so the matrix AB is defined and will be of order 2×2. Find the entry in the first row, second column of AB.

$$A = \begin{bmatrix} 4 & -5 & 2 \\ -3 & 8 & 7 \end{bmatrix}, \qquad B = \begin{bmatrix} -6 & 9 \\ 0 & -4 \\ 1 & 10 \end{bmatrix}$$

$$AB_{12} = 4(\_\_\_\_) + (\_\_\_\_)(-4) + (\_\_\_\_)(10) = \_\_\_\_\_$$

Watch and Work

Watch the videos in this section, and then complete the following exercises.

---

1. Determine $4A - 2B$, if possible.

$$A = \begin{bmatrix} 4 & -1 \\ 3 & 7 \end{bmatrix}, \quad B = \begin{bmatrix} -9 & -8 \\ -2 & -4 \end{bmatrix}$$

2. Determine $BA + B$, if possible.

$$A = \begin{bmatrix} -1 & 8 \\ -7 & 8 \end{bmatrix}, \quad B = \begin{bmatrix} 2 & -10 \end{bmatrix}$$

Practice

Solve the following problems in the space provided.

1. Determine $A + B + C$, if possible.

$$A = \begin{bmatrix} 7 & 1 \\ 6 & 8 \\ 1 & 0 \end{bmatrix}, \quad B = \begin{bmatrix} -4 & 6 \\ 8 & 2 \\ -3 & 5 \end{bmatrix}, \quad C = \begin{bmatrix} 4 & 8 \\ -6 & 3 \\ -4 & -5 \end{bmatrix}$$

2. Determine the values of the variables that will make the equation true, if possible.

$$4\begin{bmatrix} 2x & y & 3x \end{bmatrix} + 3\begin{bmatrix} 2x & 3y & -z \end{bmatrix} = \begin{bmatrix} 98 & -104 & -45 \end{bmatrix}$$

3. Determine $(BA)C$, if possible.

$$A = \begin{bmatrix} 8 & -10 \\ -4 & -8 \end{bmatrix}, \ B = \begin{bmatrix} -2 & -9 \end{bmatrix}, \text{ and } C = \begin{bmatrix} -6 & -8 \\ 1 & -7 \end{bmatrix}$$

11.5 Inverses of Matrices

Getting Started

We know that every real number has a multiplicative inverse. For example, the multiplicative inverse of 2 is $\frac{1}{2}$ since $2\left(\frac{1}{2}\right) = 1$. In the same way, we can define multiplicative inverses for certain types of matrices. Moreover, we can use these inverse matrices to solve certain systems of equations.

Topics

1. The matrix form of a system of linear equations
2. Finding the inverse of a matrix
3. Solving systems of linear equations using matrix inverses

Learn

Build the foundation you need for this section by completing the following.

1. Write the following linear system as a matrix equation: $\begin{cases} 2x - 3y + z &= -5 \\ -4x + 7y + 6z &= 11 \\ 5x - y + 3z &= -4 \end{cases}$.

$$\begin{bmatrix} 2 & \underline{\quad} & 1 \\ \underline{\quad} & 7 & 6 \\ 5 & \underline{\quad} & \underline{\quad} \end{bmatrix} \begin{bmatrix} x \\ \underline{\quad} \\ \underline{\quad} \end{bmatrix} = \begin{bmatrix} \underline{\quad} \\ 11 \\ \underline{\quad} \end{bmatrix}$$

2. The $n \times n$ **identity matrix**, denoted I_n, is the $n \times n$ matrix consisting of ______ on the main diagonal and ______ everywhere else.

3. Let A be an $n \times n$ matrix. If there exists an $n \times n$ matrix A^{-1} such that $A^{-1}A =$ ______ and $AA^{-1} =$ ______, we call A^{-1} the **inverse** of A.

4. Let $A = \begin{bmatrix} a & b \\ c & d \end{bmatrix}$. Then $A^{-1} = \frac{1}{|A|} \begin{bmatrix} d & \underline{\quad} \\ -c & \underline{\quad} \end{bmatrix}$.

5. Solve the system by the inverse matrix method: $\begin{cases} 2x + 5y &= 1 \\ 3x + 8y &= 2 \end{cases}$.

(a) $\begin{bmatrix} 2 & \underline{\quad} \\ \underline{\quad} & 8 \end{bmatrix} \begin{bmatrix} x \\ y \end{bmatrix} = \begin{bmatrix} \underline{\quad} \\ 2 \end{bmatrix}$

(b) $\begin{bmatrix} \underline{\quad} \\ y \end{bmatrix} = A^{-1} \begin{bmatrix} 1 \\ 2 \end{bmatrix}$

(c) $\begin{bmatrix} x \\ y \end{bmatrix} = \begin{bmatrix} 8 & \underline{\quad} \\ \underline{\quad} & 2 \end{bmatrix} \begin{bmatrix} \underline{\quad} \\ 2 \end{bmatrix} = \begin{bmatrix} -1 \\ \underline{\quad} \end{bmatrix}$

Watch and Work

Watch the videos in this section, and then complete the following exercises.

---

1. Solve the system by the inverse matrix method, if possible.

$$\begin{cases} -2x + 5y &= 5 \\ 6x - 15y &= -15 \end{cases}$$

2. Solve the system by the inverse matrix method, if possible.

$$\begin{cases} -3y - 2z &= 2 \\ -x - y + 4z &= -24 \\ 4x - 2y - 17z &= 88 \end{cases}$$

Practice

Solve the following problems in the space provided.

1. Find the inverse of the matrix, if possible.

$$A = \begin{bmatrix} 6 & 7 \\ -1 & -3 \end{bmatrix}$$

2. Find the inverse of the matrix, if possible.

$$A = \begin{bmatrix} -1 & -3 & -2 \\ -4 & -11 & -10 \\ 3 & 8 & 7 \end{bmatrix}$$

3. Solve the system by the inverse matrix method, if possible.

$$\begin{cases} -2x + 3y + z &= -15 \\ x + 7y - z &= -8 \\ 3x + 4y - 2z &= 7 \end{cases}$$

11.6 Partial Fraction Decomposition

Getting Started

You have had a lot of practice at combining two rational expression into a single one, but it is also useful to be able to write one rational expression as a combination of two or more others. In this section, you will learn that process, known as **partial fraction decomposition**.

Topics

1. The pattern of partial fraction decompositions
2. Completing the partial fraction decomposition process

Learn

Build the foundation you need for this section by completing the following.

---

1. Given the proper rational function $f(x) = \dfrac{p(x)}{q(x)}$, assume $q(x)$ has been completely factored as a product of factors of the form $(ax+b)^m$ and $(ax^2 + bx + c)^n$, where a, b, and c are real numbers, m and n are positive integers, and ax^2+bx+c is irreducible over the real numbers. Then $f(x)$ can be decomposed as a sum of simpler rational functions, which are determined based on the factors of $q(x)$, as follows.

   (a) Each factor of $q(x)$ of the form $(ax + b)^m$ leads to a sum of the following form.

   $$\frac{\quad\quad}{ax+b} + \frac{A_2}{\quad\quad} + \cdots + \frac{A_m}{(ax+b)^m}$$

   (b) Each factor of $q(x)$ of the form $(ax^2 + bx + c)^n$ leads to a sum of the following form.

   $$\frac{\quad\quad}{ax^2+bx+c} + \frac{A_2x + B_2}{\quad\quad} + \cdots + \frac{A_nx + B_n}{(ax^2+bx+c)^n}$$

Watch and Work

Watch the videos in this section, and then complete the following exercises.

1. Find the partial fraction decomposition of the rational function.

$$\frac{1}{x^2+9x+8}$$

2. Find the partial fraction decomposition of the rational function.

$$\frac{x+2}{(x^2+4)(x^3-3x-10)}$$

Practice

Solve the following problems in the space provided.

1. Find the partial fraction decomposition of the rational function.

$$\frac{8}{x^2 + x - 20}$$

2. Determine if the partial fraction decomposition is true or false, that is, determine if the left and right sides of the equation are equivalent.

$$\frac{8x + 44}{x^3 + 2x^2 - 5x - 6} = \frac{2}{x + 3} + \frac{4}{x - 2} + \frac{-6}{x + 1}$$

3. Find the partial fraction decomposition of the rational function.

$$f(x) = \frac{x-1}{(x^2-4)(x^2-4x+3)}$$

11.7 Systems of Linear Inequalities and Linear Programming

There are many situations that arise where constraints are given for certain variables and a solution must be found by determining all of the values that meet those constraints. Often, we'll want to find the solution that not only satisfies the constraints, but also optimizes a value. These types of problems can be modeled by linear programming. We'll look at the methods used to set up and solve such problems in this section.

Topics

1. Systems of linear inequalities
2. Planar feasible regions
3. Linear programming in two variables

Learn

Build the foundation you need for this section by completing the following.

1. If an application imposes limitations on the variables, the set of all allowable values forms a portion of the plane called the ____________ ____________ or **region of constraint**.
2. A ____________ region is one in which all the points lie within some finite distance of the origin. An ____________ region has points that are arbitrarily far away from the origin.
3. Fill in the blanks for the steps in solving a linear programming problem.

Step 1: Identify the variables to be considered in the problem, determine all the constraints on the variables imposed by the problem, and sketch the ____________ ____________ described by the constraints.

Step 2: Determine the function that is to be either maximized or minimized. This function is called the ____________ ____________.

Step 3: Evaluate the function at each of the vertices of the feasible region and compare the values found. If the feasible region is bounded, the optimum value of the function will occur at a ____________. If the feasible region is unbounded, the optimum value of the function may not exist, but if it does it will occur at a ____________.

Watch and Work

Watch the videos in this section, and then complete the following exercises.

---

1. Find the minimum and maximum values of the function, subject to the given constraints.

$$\text{Objective Function:} f(x,y) = 8x + 4y; \quad \text{Constraints:} \begin{cases} x \geq 0 \\ y \geq 0 \\ x + y \leq 9 \end{cases}$$

2. To prevent pests, an orchard can have no more than 9 times as many apple trees as peach trees. Also, the number of apple trees plus 3 times the number of peach trees must not exceed 372. The revenue from a single apple tree is \$140 and the revenue from a single peach tree is \$365. Determine the number of each type of tree that will maximize revenue. What is the maximum revenue?

Practice

Solve the following problems in the space provided.

1. Find the minimum and maximum values of the function, subject to the given constraints.

$$\text{Objective Function: } f(x,y) = 11x + 13y; \quad \text{Constraints: } \begin{cases} x \geq 0;\ y \geq 0 \\ 11x + 3y \leq 66 \\ 11x + 3y \leq 132 \end{cases}$$

2. On your birthday your father-in-law gave you \$19,000. You would like to invest at least \$4000 of the money in municipal bonds yielding 5% and no more than \$2000 in Treasury bills yielding 6%. How much should be placed in each investment in order to maximize the interest earned in one year? Assume simple interest applies. Let x represent the amount of money in municipal bonds and y represent the amount of money in Treasury bills.

11.8 Systems of Nonlinear Equations and Inequalities

Getting Started

To this point, we have only considered systems of equations in which each equation was linear. In this section we will look at systems of equations in which the equations could be *nonlinear.* As you might expect, in this case we may have multiple points at which, say, a circle intersects a parabola. Luckily, the methods of elimination and substitution that we used to solve linear systems can also be used to solve nonlinear systems.

Topics

1. Solving systems of nonlinear equations by graphing
2. Solving systems of nonlinear equations algebraically
3. Solving systems of nonlinear inequalities

Learn

Build the foundation you need for this section by completing the following.

1. Consider the nonlinear system $\begin{cases} x - y &= 3 \\ x^2 + y^2 &= 9 \end{cases}$. The graphs of the equations are shown below. What are the points of intersection?

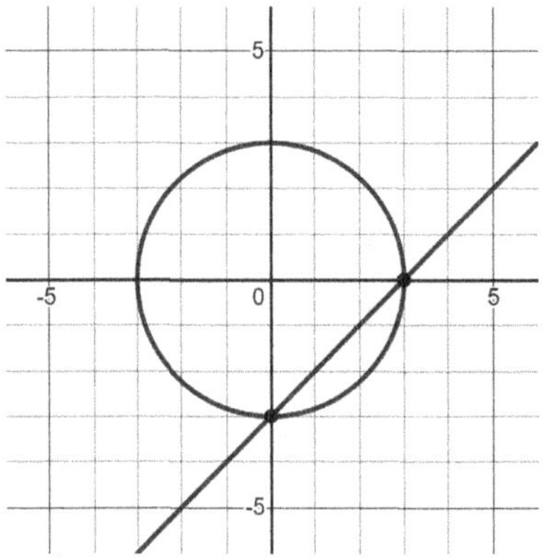

(_____,_____) and (_____,_____)

2. Consider the nonlinear system $\begin{cases} y - x^2 &= 4 \\ x^2 + y^2 &= 16 \end{cases}$. Find the y-coordinates of the solutions to this system algebraically.

   (a) Solve the first equation for x^2: $x^2 =$ ____________________

   (b) Solve the second equation for x^2: $x^2 =$ ____________________

   (c) Equate the expressions and solve for y:

$$\begin{aligned} y - 4 &= 16 - y^2 \\ y^2 + y - \_\_\_\_ &= 0 \\ (y+5)(y - \_\_\_) &= 0 \\ y &= -5, \_\_\_\_ \end{aligned}$$

3. The solutions of $y \leq -x^2 + 1$ are those points for which the y-coordinate lies on or ____________ the parabola $y = -x^2 + 1$.

Watch and Work

Watch the videos in this section, and then complete the following exercises.

---

1. Solve the nonlinear system.

$$\begin{cases} x^2 + y^2 &= 12 \\ 2x^2 + 3y^2 &= 40 \end{cases}$$

2. Carlos takes 10 minutes longer than Patricia to make the 110-mile drive between two cities. Patricia drives 6 miles an hour faster. How fast do Carlos and Patricia drive?

Practice

Solve the following problems in the space provided.

1. Solve the nonlinear system.

$$\begin{cases} x^2 + y^2 &= 61 \\ 2x^2 - 4y^2 &= -28 \end{cases}$$

2. Solve the nonlinear system

$$\begin{cases} x^2 + y^2 &= 8 \\ y - 2 &= -x^2 \end{cases}$$

3. The product of two positive integers is 45, and their sum is 14. What are the integers?

Chapter 12

Sequences, Series, Combinatorics, and Probability

As the title suggests, in this chapter we will cover a wide variety of topics. Sequences and series are used extensively in mathematics and a solid working knowledge of these topics will be very beneficial to future mathematical studies. Probability and combinatorics form a foundation for the further study of statistics. We will be introduced to all of these topics in this chapter.

12.1 Sequences and Series

Getting Started

Sequences occur naturally in a wide variety of areas. Recognizing sequences and finding convenient ways to express them are skills that we will learn in this section. In addition, we'll investigate what happens when we add the entries in a sequence together. These are called series.

Topics

1. Recursively and explicitly defined sequences
2. Summation notation and formulas
3. Partial sums and series
4. Fibonacci sequences

Learn

Build the foundation you need for this section by completing the following.

---

1. Determine the first three terms of the sequence $a_n = 2n + 3$.

   (a) $a_1 = 2(1) + 3 =$ ________

   (b) $a_2 = 2($______$) + 3 =$ ________

   (c) $a_3 = 2($______$) + 3 =$ ________

2. A ________________ formula is one that refers to one or more of the terms preceding a_n in the definition for a_n.

3. The sum $a_1 + a_2 + \cdots + a_n$ is expressed in summation notation as ______________

4. An infinite series $\sum_{i=1}^{\infty} a_i$ ______________ if the sequence of partial sums $S_n = \sum_{i=1}^{n} a_i$ approaches a fixed real number S. If the sequence of partial sums does not approach some fixed number, we say the series ______________.

Watch and Work

Watch the videos in this section, and then complete the following exercises.

1. Determine the first five terms of the sequence whose n^{th} term is defined as follows.

$$a_1 = 2 \text{ and } a_n = \sqrt{(-2a_{n-1})^2 - 2} \text{ for } n \geq 2$$

2. Translate the summation notation that follows into an expanded sum. Then evaluate the sums.

$$\sum_{i=3}^{7}(i-5)$$

Practice

Solve the following problems in the space provided.

1. Determine the first five terms of the sequence whose n^{th} term is defined as follows.

$$a_n = \frac{-2n^2}{-3n}$$

2. Find a possible formula for the general n^{th} term of the sequence that begins as follows.

$$-3, 9, -27, 81, -243, \ldots$$

3. Find a formula for the nth partial sum S_n of the series. If the series is finite, determine the sum. If the series is infinite, determine if it converges or diverges, and if it converges, determine the sum.

$$\sum_{i=1}^{20} \left((-2)^i - (-2)^{i-1}\right)$$

12.2 Arithmetic Sequences and Series

Getting Started

An *arithmetic* sequence is one of the more common types of sequences. An arithmetic sequence is simply one in which each pair of successive terms has the same difference. We'll investigate these types of sequences as well as their associated series in this section.

Topics

1. Arithmetic sequences and series
2. The formula for the general term of an arithmetic sequence
3. Evaluating partial sums of arithmetic sequences

Learn

Build the foundation you need for this section by completing the following.

1. A sequence $\{a_n\}$ is an **arithmetic sequence** if there is a constant d so that _______ − _______ $= d$ for each $n = 1, 2, 3, \ldots$.

2. Complete the formula for the partial sum of an arithmetic sequence $\{a_n\}$.

$$S_n = na_1 + d\sum_{i=1}^{n-1} i = na_1 + d\left(\underline{\qquad\qquad}\right)$$

3. Find the formula for the general n^{th} term of the arithmetic sequence $-3, 2, 7, 12, \ldots$.

$$a_n = a_1 + d(n-1) = \underline{\qquad} + \underline{\qquad}(n-1) = \underline{\qquad\qquad\qquad}$$

4. Use the formula $S_n = \left(\frac{n}{2}\right)(a_1 + a_n)$ to determine S_5 for the arithmetic sequence $3, 7, 11, 15, 19, \ldots$.

$$S_5 = \left(\frac{5}{2}\right)(\underline{\qquad} + \underline{\qquad}) = \left(\frac{5}{2}\right)(\underline{\qquad}) = \underline{\qquad\qquad\qquad}$$

Watch and Work

Watch the videos in this section, and then complete the following exercises.

1. Find the explicit formula for the general n^{th} term of the arithmetic sequence described below.

$$a_{16} = \frac{79}{2} \text{ and } d = \frac{5}{2}$$

2. The following is a partial sum of an arithmetic sequence; use either formula to determine its value.

$$4 + 12 + \cdots + 164$$

Practice

Solve the following problems in the space provided.

1. Find the explicit formula for the general n^{th} term of the arithmetic sequence described below.

$$a_{28} = -48 \text{ and } a_{36} = -64$$

2. Given that $\{a_n\}$ is an arithmetic sequence, $a_1 = -9$ and $a_5 = 11$, what is a_{75}?

3. A person accepts a position with a company at a salary of $28,000 for the first year. The person is guaranteed a raise of $1700 per year for the first 4 years. Determine the person's total compensation through 5 full years of employment.

12.3 Geometric Sequences and Series

Getting Started

Another common type of sequence is called a *geometric* sequence. In this section, we'll investigate the properties of these sequences and look at the conditions that will cause the associated infinite geometric series to converge to a finite number.

Topics

1. Geometric sequences
2. The formula for the general term of a geometric sequence
3. Evaluating partial sums of geometric sequences
4. Evaluating infinite geometric series
5. Zeno's paradoxes

Learn

Build the foundation you need for this section by completing the following.

1. A sequence $\{a_n\}$ is a **geometric sequence** if there is some constant $r \neq 0$ so that $\overline{} = r$ for each $n = 1, 2, 3, \ldots$.

2. Determine the formula for the general n^{th} term of the geometric sequence $2, -6, 18, \ldots$.

   (a) $r = \dfrac{-6}{2} = \_\_\_\_$

   (b) $a_1 = \_\_\_\_$

   (c) $a_n = a_1 r^{n-1} = \_\_\_\_(\_\_\_\_)^{n-1}$

3. Find the sum of the first 6 terms of the geometric sequence $1, -2, 4, -8, \ldots$.

$$S_6 = \frac{a_1(1 - r^6)}{1 - r} = \frac{1(1 - (\_\_\_\_)^6)}{1 - (\_\_\_\_)} = \frac{1(1 - \_\_\_\_)}{1 - (\_\_\_\_)} = \_\_\_\_\_\_ = \_\_\_\_$$

4. If $|r| < 1$, the infinite geometric series $\sum_{n=1}^{\infty} a_1 r^{n-1} = a_1 + a_1 r + a_1 r^2 + a_1 r^3 + \cdots$ converges, and the sum of the series is given by $S =$ ______.

5. Find the sum of the infinite geometric series $\sum_{n=1}^{\infty} 2\left(\frac{1}{3}\right)^{n-1}$.

$$\sum_{n=1}^{\infty} 2\left(\frac{1}{3}\right)^{n-1} = \frac{\rule{2em}{0.4pt}}{1-\frac{1}{3}} = \frac{\rule{2em}{0.4pt}}{\rule{2em}{0.4pt}} = \rule{4em}{0.4pt}$$

Watch and Work

Watch the videos in this section, and then complete the following exercises.

---

1. Given that $\{a_n\}$ is a geometric sequence, $a_1 = 1$, and $a_3 = \frac{1}{4}$, what is a_{12}?

2. Determine if the infinite geometric series converges. If so, find the sum.

$$\sum_{i=0}^{\infty} 2\left(\frac{-3}{4}\right)^i$$

Practice

Solve the following problems in the space provided.

1. Find the explicit formula for the general n^{th} term of the geometric sequence.

$$a_6 = \frac{46875}{2} \text{ and } r = 5$$

2. The following sum is a partial sum of a geometric sequence. Use this fact to evaluate the sum.

$$-1 + (-3) + \cdots + (-6561)$$

3. A rubber ball is dropped from a height of 81 feet, and on each bounce it rebounds up 48% of its previous height. How far has it traveled vertically at the moment when it hits the ground for the 22nd time?

12.4 Mathematical Induction

Getting Started

Often times, mathematicians want to prove that a certain statement is true. When this statement is to apply to all of the natural numbers, one of the most common methods of proof is *proof by induction.* In this section, we'll look at the general methodology of mathematical induction and use it to prove some basic propositions about the natural numbers.

Topics

1. The role of induction
2. Proofs by mathematical induction

Learn

Build the foundation you need for this section by completing the following.

1. Assume that $P(n)$ is a statement about each natural number n. Suppose that the following two conditions hold:

   1. $P(\_\_\_)$ is true.
   2. For each natural number k, if $P(k)$ is true then $P(\_\_\_\_\_)$ is true.

   Then the statement $P(n)$ is true for every natural number n.

2. Consider the following statement $P(n)$: $1+2+4+8+\cdots+2^{n-1}=2^n-1$.

   (a) Show that $P(1)$ is true: $1 = \_\_\_\_ - 1$

   (b) Write the statement $P(k)$: $1+2+\cdots+\_\_\_\_\_ = \_\_\_\_ - 1$

   (c) Write the statement $P(k+1)$: $1+2+\cdots+\_\_\_\_\_\_ = \_\_\_\_\_ - 1$

Watch and Work

Watch the videos in this section, and then complete the following exercises.

1. Find S_{k+1} for the given S_k.

$$S_k = \frac{k}{(k+1)(k-1)}$$

2. Use the Principle of Mathematical Induction to prove the statement.

$$1(1!) + 2(2!) + \cdots + n(n!) = (n+1)! - 1$$

Practice

Solve the following problems in the space provided.

1. Use the Principle of Mathematical Induction to prove the statement.

$$\frac{1}{1\cdot 2}+\frac{1}{2\cdot 3}+\frac{1}{3\cdot 4}+\cdots+\frac{1}{n(n+1)}=\frac{n}{n+1}$$

2. Use the Principle of Mathematical Induction to prove the statement.

$$\frac{1}{2}+\frac{1}{4}+\frac{1}{8}+\cdots+\frac{1}{2^n}=\frac{2^n-1}{2^n}$$

12.5 Combinatorics

Getting Started

In this section, we'll look at the basics of combinatorics. The general purpose of combinatorics is to be able to count complex things quickly and easily. This skill should not be underestimated. We'll look at the basics of combinatorics including permutations and combinations, and then we'll investigate a couple of important theorems.

Topics

1. The Multiplication Principle of Counting
2. Permutations
3. Combinations
4. The Binomial Theorem
5. The Multinomial Theorem

Learn

Build the foundation you need for this section by completing the following.

1. Suppose $E_1, E_2, \ldots, E_n$ is a sequence of events, each of which has a certain number of possible outcomes. Suppose event E_1 has m_1 possible outcomes, and that after event E_1 has occurred, event E_2 has m_2 possible outcomes. Similarly, after event E_2, event E_3 has m_3 possible outcomes, and so on. Then the total number of ways that all n events can occur is ____________________.

2. The number of permutations of n objects taken k at a time is ${}_nP_k = \frac{n!}{\_\_\_\_\_\_}$.

3. The number of combinations of n objects taken k at a time is ${}_nC_k = \frac{n!}{\_\_\_\_\_\_}$.

4. The expression $\binom{n}{k}$ is called a ________________ ________________, as it corresponds to the coefficient of $x^k y^{n-k}$ in the expansion of $(x+y)^n$.

5. The Multinomial Theorem: Given the positive integer n and expression $A_1, A_2, \ldots, A_r$,

$$(A_1 + A_2 + \cdots + A_r)^n = \sum_{k_1+k_2+\cdots+k_r=n} \binom{n}{k_1, k_2, \ldots, k_r} \rule{4cm}{0.4pt}$$

Watch and Work

Watch the videos in this section, and then complete the following exercises.

---

1. A license plate must contain three numerical digits followed by three letters. if the first digit cannot be 0 or 1, how many different license plates can be created?

2. Suppose you have 10 Physics texts, 8 Computer Science texts, and 12 Math texts. How many different ways can you select 2 of each to take with you on vacation?

Practice

Solve the following problems in the space provided.

1. Expand the expression $(5x + y)^5$.

2. A trade union asks its members to select 3 people from a slate of 10 to serve as representatives at a national meeting. How many different sets of 3 can be chosen?

3. How many ways are there of choosing three cards from a standard 52-card deck and arranging them in a row? How many different three-card hands can be dealt from a standard 52-card deck?

12.6 Probability

Getting Started

The mathematical field of probability is one of the most applicable to real-world situations. It is used in everything from gambling to insurance. In this section, we'll introduce some of the basic language and concepts of this important topic.

Topics

1. The language of probability
2. Computing probabilities using combinatorics
3. Unions, intersections, and independent events

Learn

Build the foundation you need for this section by completing the following.

1. An **experiment** is any activity that results in well-defined ____________.

2. If E is an event of an experiment with equally likely outcomes, and if S is the sample space of the experiment, then the probability of E is given by $P(E) = \dfrac{\quad}{n(S)}$.

3. Let $E = \{1, 3, 5\}$ and $F = \{1, 2\}$. Find the following.

   (a) $E \cup F =$ ____________

   (b) $E \cap F =$ ____________

4. If $E_1, E_2, \ldots, E_n$ are pairwise disjoint, then

$$P(E_1 \cup E_2 \cup \cdots \cup E_n) = \underline{\qquad} + \underline{\qquad} + \cdots + \underline{\qquad}$$

Watch and Work

Watch the videos in this section, and then complete the following exercises.

1. A card is drawn from a standard 52-card deck. Find the probability of drawing a face card (Jack, Queen, or King) in the suit of Spades.

2. A coin is flipped 5 times. Find the probability of getting tails exactly 4 times.

Practice

Solve the following problems in the space provided.

1. Determine the sample space of the experiment.

   A slot machine lever is pulled; there are 5 slots, each of which can hold 5 different values.

2. What is the probability that a 7-digit driver's license number chosen at random will not have 8 as a digit?

3. A pair of dice is rolled. Find the probability that the sum of the two dice is 5 or 8.

Chapter 13

An Introduction to Limits, Continuity, and the Derivative

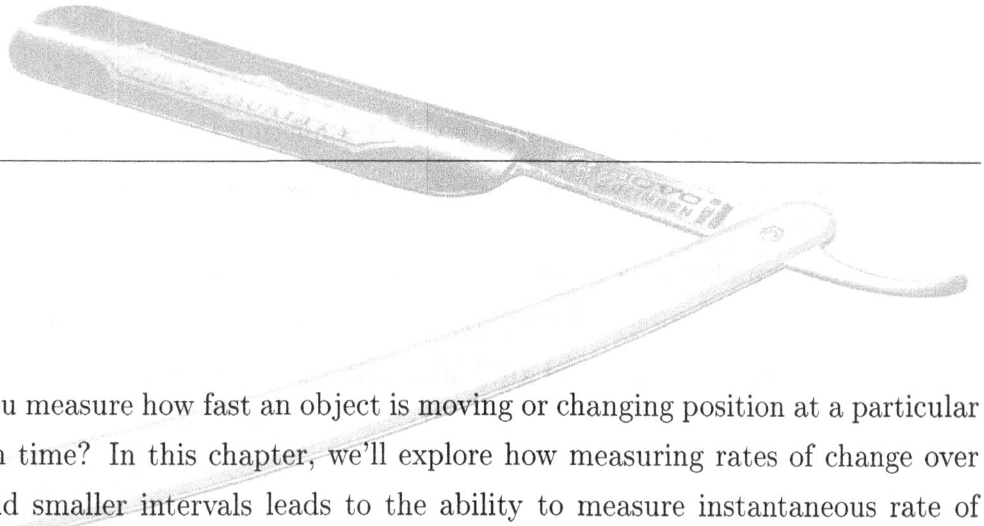

How do you measure how fast an object is moving or changing position at a particular moment in time? In this chapter, we'll explore how measuring rates of change over smaller and smaller intervals leads to the ability to measure instantaneous rate of change. This will be done using derivatives, one of the most important concepts in calculus and all of mathematics.

13.1 Rates of Change and Tangents

Getting Started

To measure a rate of change, we measure the change in position over the corresponding change in time. This concept is foundational for the study of motion. We will use the concept of rate of change to look at velocity as well as how to identify the slope of a tangent line for the graph of a function.

Topics

1. The difference quotient
2. The velocity problem
3. The tangent problem

Learn

Build the foundation you need for this section by completing the following.

1. Given a real number c, a nonzero real number h (which may be positive or negative), and a function f defined on an interval containing the points c and $c+h$, the **difference quotient of f at c with increment h** is the ratio

$$\frac{f(\qquad) - f(\quad)}{h}.$$

2. The average rate of change of f over the interval with endpoints c and $c+h$ represents the slope of the ___________ __________ drawn between the points $(c, f(c))$ and $(c+h, f(c+h))$ on the graph of f.

3. The **average rate** r of travel for an object moving distance d over time t is the ____________ of d over t.

4. In mathematics, a line is tangent to a given curve if it "_________ ____________" the curve.

Watch and Work

Watch the videos in this section, and then complete the following exercises.

1. Use the difference quotient of the function for one appropriate value of c to determine the average rate of change of the function over each of the given intervals.

$$f(x) = x^2 - 2x; \quad \textbf{a.}\ [4,5] \quad \textbf{b.}\ [4,4.5] \quad \textbf{c.}\ [3.9,4]$$

2. Use difference quotients to approximate the slope of the tangent to the graph of the function at the given point. Use at least five different h-values that are decreasing in magnitude.

$$g(x) = \sin(x); (0,0)$$

Practice

Solve the following problems in the space provided.

1. For the function below, construct and simplify the difference quotient at c with increment h.

$$f(x) = \frac{1}{2-x}$$

2. The distance, in meters, traveled by a particle moving through space in t seconds is given by $d(t) = 2t^2 - t$. Estimate the instantaneous velocity at $t = 2$ seconds.

3. Use difference quotients to approximate the slope of the tangent to the graph of the function at the given point. Use at least five different h-values that are decreasing in magnitude.

$$g(x) = x^3 - 3x; (1, -2)$$

13.2 Limits in the Plane

Getting Started

How do you describe a situation where a variable is getting closer and closer to a value, but never actually getting there? You use limits. In this section, you will learn how mathematicians define the concept of limits and how they are used to describe the graphs of certain functions.

Topics

1. Limits in verbal, numerical, and visual forms
2. Vertical asymptotes and one-sided limits
3. Horizontal asymptotes and limits at infinity

Learn

Build the foundation you need for this section by completing the following.

1. We say that the limit of $f(x)$ is L as x approaches c, and write

$$\lim_{x \to c} f(x) = L,$$

if the value of $f(x)$ is ______________ and ______________ to L as x takes on values closer and closer (but not equal to) c.

2. If $\lim_{x \to c} f(x) = \infty$, then we say that f has a ______________ ______________ at the point c.

3. We say that the line $y = L$ is a ______________ ______________ for the function f if either of the following statements is true.

$$\lim_{x \to \infty} f(x) = L \quad \text{or} \quad \lim_{x \to -\infty} f(x) = L$$

Watch and Work

Watch the videos in this section, and then complete the following exercises.

1. Create a table of values to estimate the value of the indicated limit without graphing the function.

$$\lim_{x \to 5} \frac{2x^2 - 8}{x - 2}$$

2. Use one-sided limit notation to describe the behavior of the function near its vertical asymptote(s).

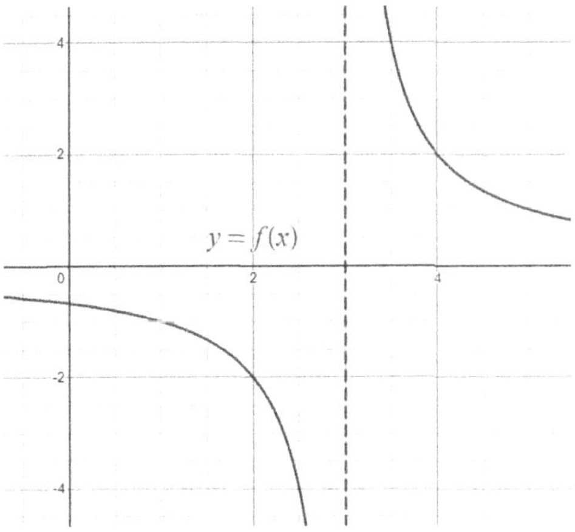

Practice

Solve the following problems in the space provided.

1. Create a table of values to estimate the value of the indicated limit without graphing the function.

$$\lim_{x \to 0} \frac{2 \sin x}{5x}$$

2. Use the graph to find the indicated one-sided limits, if they exist.

a. $\lim_{x \to 2^+} g(x)$ b. $\lim_{x \to 2^-} g(x)$

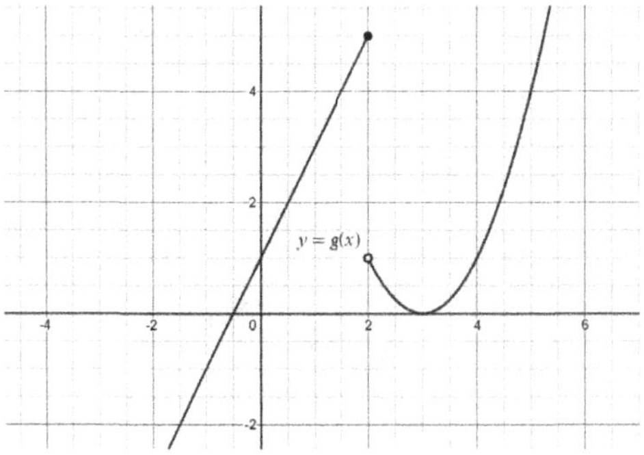

3. Use limit notation to describe the unbounded behavior of the given function as x approaches ∞ and/or $-\infty$.

$$f(x) = -x^3 + 3x - 1$$

13.3 The Mathematical Definition of Limit

Getting Started

Suppose you had a pet frog and you placed it at the end of a table that was one meter long. Every time you clapped, the frog jumped halfway to the end of the table. After one clap, he has jumped 0.5 m, after two claps, 0.75 m, and so on. Technically, the frog will never get to the end of the table. However, if you kept on clapping, the distance the frog has jumped will get as close as you like to one meter. In other words, you could say that the **limit** of the distance that the frog jumps is one meter. You'll study the concept of limit in a slightly more formal way in this section, but feel free to bring your frog along.

Topics

1. The formal definition of a limit
2. Proving a limit exists
3. Proving a limit does not exist

Learn

Build the foundation you need for this section by completing the following.

---

1. Let f be a function defined on an open interval containing c, except possibly at c itself. We say that the limit of $f(x)$ as x approaches c is L, and write $\lim_{x \to c} f(x) = L$, if for every number $\varepsilon > 0$ there is a number $\delta > 0$ such that $|$________________$| < \varepsilon$ whenever $0 < |$____________$| < \delta$.

2. Let f be a function defined on an open interval containing c, except possibly at c itself. We say that the limit of $f(x)$ as x approaches c is positive infinity, and write $\lim_{x \to c} f(x) = \infty$, if for every positive number M there is a number $\delta > 0$ such that ______________ whenever $0 < |$____________$| < \delta$.

3. Let f be a function defined on some interval (a, ∞). We say that the limit of f at infinity is L, and write $\lim_{x \to \infty} f(x) = L$, if for every number $\varepsilon > 0$ there is a number N such that $|$________________$| < \varepsilon$ for all ____________.

Watch and Work

Watch the videos in this section, and then complete the following exercises.

1. Find a $\delta > 0$ that satisfies the limit claim corresponding to $\varepsilon = 0.1$, that is, such that $0 < |x - c| < \delta$ would imply $|f(x) - L| < 0.1$.

$$\lim_{x \to 3}(2x - 4) = 8$$

2. For the given function $f(x)$, find a $\delta > 0$ corresponding to $M = 100$, that is, such that $0 < |x - c| < \delta$ would imply $f(x) > 100$.

$$\lim_{x \to 2} \frac{1}{(x-2)^2}$$

Practice

Solve the following problems in the space provided.

1. Find a $\delta > 0$ that satisfies the limit claim corresponding to $\varepsilon = 0.1$, that is, such that $0 < |x - c| < \delta$ would imply $|f(x) - L| < 0.1$.

$$\lim_{x \to 1} (x^2 - 1) = 0$$

2. Find a number N that satisfies the limit claim corresponding to $\varepsilon = 0.1$, that is, such that $x > N$ (or $x < N$, as appropriate) would imply $|f(x) - L| < 0.1$.

$$\lim_{x \to \infty} \frac{x}{2x + 1} = \frac{1}{2}$$

3. Use the $\varepsilon - \delta$ definition to prove the limit claim.

$$\lim_{x \to 2}(4x - 1) = 7$$

13.4 Determining Limits of Functions

Getting Started

Now that we have the feel for what a limit is, we can go a little deeper. Indeed, beyond just proving limit claims, we would like to be able to determine the limits of a variety of functions. To do this, we'll need to learn the limit laws which tell us how to apply some basic operations to the concept of limits.

Topics

1. Limit laws
2. Limit determination techniques

Learn

Build the foundation you need for this section by completing the following.

1. Let f and g be two functions such that both $\lim_{x \to c} f(x)$ and $\lim_{x\to c} g(x)$ exist, and let k be a fixed real number. Then the following laws hold.

   (a) $\lim_{x \to c} [f(x) + g(x)] =$ ____________ + ____________

   (b) $\lim_{x \to c} [kf(X)] =$ ____________

   (c) $\lim_{x \to c} \dfrac{f(x)}{g(x)} =$ ____________, provided $\lim_{x \to c} g(x) \neq 0$

2. Let p be a polynomial function. Then

$$\lim_{x \to c} p(x) = \_\_\_\_\_\_\_\_\_\_.$$

3. If $g(x) \leq f(x) \leq h(x)$ for all x in some open interval containing c, except possibly at c itself, and if $\lim_{x \to c} g(x) = \lim_{x \to c} h(x) = L$, then ________________ as well.

4. If $f(x) \leq g(x)$ for all x in some open interval containing c, except possibly at c itself, and if the limits of f and g both exist at c, then

$$\_\_\_\_\_\_\_\_\_\_ \leq \_\_\_\_\_\_\_\_\_\_.$$

Watch and Work

Watch the videos in this section, and then complete the following exercises.

1. Use appropriate limit laws to evaluate the given limit.

$$\lim_{x \to 2} (x^2 - 3x + 1)$$

2. Use algebra to evaluate the given limit.

$$\lim_{x \to 5} \frac{x^2 - 25}{x - 5}$$

Practice

Solve the following problems in the space provided.

---

1. Use appropriate limit laws to evaluate the given limit.

$$\lim_{x \to 1} \frac{2x + 4}{x^2 - 2}$$

2. Use algebra to evaluate the given limit.

$$\lim_{x \to 0} \frac{\sqrt{x + 3} - \sqrt{3}}{x}$$

3. Use algebra to evaluate the given limit.

$$\text{If } f(x) = 4 - x^2, \text{ find } \lim_{h \to 0} \frac{f(1+h) - f(1)}{h}.$$

13.5 Continuity

Getting Started

Continuity is a fairly intuitive concept to grasp. One way to think about it is that a function is continuous if the graph can be drawn without picking up your pencil (or chalk, for that matter). But as is the case for many things in mathematics, we don't want to rely on intuition alone. In this section, we'll look at the formal definition of continuity and investigate some of its properties.

Topics

1. Continuity and discontinuity at a point
2. Continuous functions
3. Properties of continuity

Learn

Build the foundation you need for this section by completing the following.

1. Given a function f defined on an open interval containing c, we say f is continuous at c if

$$\lim_{x \to c} f(x) = \underline{\hspace{3cm}}.$$

2. If a function f has a point of discontinuity at c but $\lim_{x \to c} f(x)$ exists, c is called a ____________ ____________ of f.

3. Given a function f and a point c in the domain of f, we say f is continuous at c if for every number $\varepsilon > 0$ there is a number $\delta > 0$ such that $|\underline{\hspace{3cm}}| < \varepsilon$ for all x in the domain of f satisfying $0 < |\underline{\hspace{3cm}}| < \delta$.

4. A function is said to be continuous if it is continuous at every point of its ____________.

5. If f is a continuous function defined on the closed interval $[a, b]$, the f takes on every value between $f(a)$ and __________.

Watch and Work

Watch the videos in this section, and then complete the following exercises.

1. Find and classify the discontinuities (if any) of the function as removable or nonremovable.

$$f(x) = \frac{x^2 - 3x}{x^2 - 4x + 3}$$

2. Decide whether the Intermediate Value Theorem applies to the given function on the indicated interval. If so, find c as guaranteed by the theorem. If no, find the reason.

$$f(x) = x^2 - 2x - 3 \text{ on } [1, 4]; \quad f(c) = 0$$

Practice

Solve the following problems in the space provided.

1. Find and classify the discontinuities (if any) of the function as removable or nonremovable.

$$f(x) = \frac{x+4}{x^2+2x-8}$$

2. Use the $\epsilon - \delta$ definition to prove that the function is continuous.

$$f(x) = 4x - 1$$

3. Use the Intermediate Value Theorem to prove that the given equation has a solution on the indicated interval.

$$x^3 - 3x^2 + 1.5 = 0 \text{ on } [-1, 3]$$

13.6 The Derivative

Getting Started

This chapter culminates with one of the most important concepts in calculus: the derivative. We now have everything that we need to precisely define and calculate the instantaneous velocity of an object at a given time. This is exactly what the derivative measures. In this section, we'll learn how to find the derivative of a given function and what the derivative can tell us about that function.

Topics

1. The velocity and tangent problems revisited
2. The derivative as a function

Learn

Build the foundation you need for this section by completing the following.

1. The derivative of the function f at the point c, denoted $f'(c)$, is

$$f'(c) = \lim_{h \to 0} \frac{f(\qquad\qquad) - f(c)}{h},$$

provided the limit exists.

2. The derivative of the function f at the point c, $f'(c)$, indicates the __________ of change of the function f at the point c.

3. If $v(t)$ represents the velocity of an object at time t, $s(t) =$ ______ represents the speed at time t.

4. If the cost function $C(x)$ represents the total cost of producing x items of a certain product, then the derivative $C'(x)$ is called the __________ cost.

Watch and Work

Watch the videos in this section, and then complete the following exercises.

1. Find the equation of the tangent line to the graph of $f(x)$ at the given point.

$$f(x) = 2x^2 - 3; \quad (-1, -5)$$

2. Use the definition to find the derivative function f' of the given function f. Find all x-values (if any) where the tangent line is horizontal.

$$f(x) = 2x - 3$$

Practice

Solve the following problems in the space provided.

1. Use the definition to find the derivative function f' of the given function f. Find all x-values (if any) where the tangent line is horizontal.

$$f(x) = x^2 - 3x + 1$$

2. Find the equation of a tangent line to the graph of the function that is parallel to the given line.

$$f(x) = 2 - x^2; \quad x + y = 2$$

3. Use the alternate form of the definition of the derivative $f'(c) = \lim_{x \to c} \frac{f(x) - f(x)}{x - c}$ to evaluate the given slope.

$$f(x) = \frac{1}{x+2}; \quad f'(0)$$

Notes

Notes

Notes

Notes

Notes

Notes

Notes

Notes

Notes